RAND

An Optical Signal Processing Model for the Interferometric Fiber Optic Gyro

Volume 1, Deterministic Model

Joseph M. Aein

Prepared for the
Advanced Research Projects Agency

National Security Research Division

This report presents a mathematical model for analyzing the optical signal processing employed by closed-loop interferometric fiber optic gyroscopes (IFOG) used in high-performance (navigation-grade) solid-state miniature inertial measurement units (MIMU). Development of a highly accurate, low-cost MIMU is a central objective of the Advanced Research Project Agency's effort in range-invariant guidance technologies. In combination with a miniature Global Positioning System (GPS) receiver (MGR), the MIMU forms a GPS-based guidance package (GGP) suitable for insertion into a wide variety of Department of Defense platforms.

The results of RAND's effort are being published in two volumes: The present volume provides the necessary theory to analyze the desired "sure" signal operation of the IFOG and some of the unavoidable distortion terms. Volume 2 will provide the stochastic theory necessary to analyze the randomly scattered "noise" light produced when the desired optical signals impinge on microimperfections distributed throughout the fiber waveguides in the IFOG.

This publication is one of several RAND reports that track the progress of these new guidance technologies. The RAND study has been supported by the Advanced Systems Technology Office (ASTO) of ARPA for several years, and RAND's effort is continuing as ARPA completes the brassboard fabrication phase of the GGP and commences the next phase, which is to produce GGP military demonstration units. The present work was carried out in the Applied Science and Technology Program of the National Defense Research Institute, RAND's federally funded research and development center sponsored by the Office of the Secretary of Defense, the Joint Staff, and the defense agencies.

CONTENTS

Appendix

FIGURES

This report presents the results of a communication theoretic model used to analyze the operation of interferometric fiber optic gyroscopes (IFOG). The IFOG is an all-solid-state rotation-rate sensor employed in miniature inertial measurement units (MIMU). With careful design and fabrication the IFOG can achieve a navigation grade of performance (0.01 deg/hr drift rate). The IFOG offers the considerable advantage of reducing instrument costs by a factor of two or more over the current ring laser gyroscope (RLG) technology base.

Inertially derived motion data from a MIMU can be combined with radiometric position data from a miniature Global Positioning System (GPS) receiver (MGR) to provide a low-cost, highly adaptable GPS/inertial navigation system (GPS/INS). GPS/INS units are able to operate through periods of GPS signal loss, perform autonomous in-flight inertial alignment, and track GPS through higher dynamic g-loads.

This report analyzes the optical signal processing used to achieve high-performance IFOG rotation sensors. A simplified classical propagation model is used to describe the input/output relationships of the optical beams employed. Optical signal modulations are impressed with the two following goals:

1. Increase the linearity of the rotation-rate measurement and extend the range of rotation rates over which the instrument is useful (dynamic range); and

2. Mitigate the deleterious effects of the optical scatter noise unavoidably generated by the desired light beams as they traverse the optical pathways of the IFOG.

This report establishes the model and analytic framework to be used and then deals with the goal (1), above. A companion volume will deal with goal (2).

In the deterministic theory employed here, the following are accomplished: (1) AC bias modulation used to linearize the IFOG instrument is analyzed; (2) closed-loop, feedback IFOG operation is described to extend the dynamic range; and (3) ideal serrodyne operation is explained. In an appendix, a detailed analysis is presented for the distortions incurred with real (i.e., nonideal) serrodyne operation. The analysis establishes the performance relationship among the appropriate optical signal parameters and allows for their optimization.

The feasibility of constructing navigation-grade IFOG rotation-rate sensors has been demonstrated recently by ARPA. The principal IFOG effort now under way at ARPA is directed toward realizing reduced production costs afforded by IFOG technology. This includes IFOG design employing lower-cost components as well as reducing instrument assembly and test costs through automation.

ACKNOWLEDGMENTS

The author wishes to express his appreciation and gratitude to those assisting him in bringing this report to fruition:

To Dr. Larry Stotts and Major Beth Kaspar of ARPA's Advanced Systems Technology Office for their unflagging encouragement tempered by a virtual residence in Missouri.

To Messrs. Charles Bass and Matt McLandrich of the Naval Command, Control and Ocean Surveillance Center (NCCOSC) RDT&E Division (NRaD) for their attention to the physical details of the analysis.

To Dr. Richard Greenspan of the Charles Stark Draper Laboratories, and Drs. Edward Bedrosian and Phillip Feldman of RAND for reviewing the draft manuscripts. No line of argument or turn of phrase was left unimproved.

To Patricia Bedrosian for editing this report into a style of English that the author would just as soon have the reader believe was his own.

To Dr. Sherman Karp, who said to the author "it wasn't so," thereby necessitating the whole endeavor. And it wasn't so.

Chapter One

INTRODUCTION

OVERVIEW

This report develops the elements of an optical signal processing model useful for analyzing the performance of Interferometric Fiber Optic Gyros (IFOG). The IFOG is used as an inertial rotation-rate sensor; hence, the use of the term "gyro." However, it has no moving parts because it senses angular rotation rate by means of the Sagnac effect.[1] A simplified classical propagation description of the guided light waves is employed to develop the desired signal processing model. By analogy with radio frequency (RF) modeling (e.g., delay lines, the multipath channel, and radar clutter), a representation is obtained suitable for characterizing both (1) the deterministic processing properties for the desired optical signal(s) and (2) the optical-signal-induced scatter interference that corrupts the desired signals.

The key issue of interest here is to quantitatively model the optical modulation processes that can be employed to (1) extend the range of input rotation rates over which the IFOG remains linear (i.e., the dynamic range) and (2) mitigate the negative effects of the optical scatter noise unavoidably generated in the optical pathways of the IFOG instrument. The model employed in both Volumes 1 and 2 assumes strong optical signal waves wherein the usual, optical noise sources (e.g., photon shot noise, detector quantum noise, dark cur-

---

[1]The well-known Ring Laser Gyro (RLG) also is based on the Sagnac effect. Interest in the IFOG centers on achieving performance comparable to the RLG (< 0.01 deg/hr drift rate) at considerably less cost.

rent noise, and amplifier noise) are negligible compared to the undesired scatter light, which is the dominant noise source in these instruments. The path-generated scattered signal light is treated as the only source of interference or "noise." Results for the conventional optical noise sources already exist in the literature (Karp and Gagliardi, 1976). Since they are statistically independent noise processes, their vector sum can be "added in" to the results for the scatter light in a relatively straightforward way.

This report develops an overall model for IFOG signal processing and delineates the deterministic error waveforms resulting from the nonlinear elements of the system. Using the basic methodology developed here, a companion volume will statistically analyze the error signals caused by the desired optical signals impinging upon random, small but unavoidable optical scattering phenomena distributed throughout the fiber propagation path around the rotation sensing coil.

First, the physical operating principles of an IFOG are described and summarized.[2] The IFOG is modeled as a symmetric pair of classical communication[3] channels sharing a common physical channel albeit in opposite directions of propagation. Simple equations for the propagating optical signals of interest with their modulation format are presented. The communication theoretic results analogous to FM theory and phase lock loops are derived. The report concludes with a summary of a waveform error analysis (contained in the appendixes).

We conclude this overview with a word about references. The applied physics literature is replete with papers on describing and quantifying the physical phenomena of the Sagnac effect and the various unwanted degradations. A useful summary and an extensive bibliography first was provided by Ezekiel and Arditty (1982), but their paper involves rather advanced mathematics. The most up-to-

---

[2]A related RAND report (Aein, 1992) gives a general discussion of the IFOG as a rotation-rate sensor for the inertial measurement unit (IMU), summarizes the current state of IFOG technology, and provides some applications, especially in combination with Global Positioning System (GPS) receivers.

[3]This is a misnomer as the "transmitter" and "receiver" are collocated. More precisely, the problem at hand is one of channel *estimation*.

date review of the physical phenomena was published while this report was in review (Burns, 1994). It has the most current bibliography and was written for those having more conventional mathematical training. To this author's knowledge it is the first in print to use the term "signal processing" for describing the methods for mitigating or improving IFOG performance (Chapter 3, pp. 81–113). This, of course, is precisely the subject of this report.

This report is primarily of pedagogic value to prepare the reader for Volume 2. Brought forth here are the strong connections with conventional communications theory and its rich body of methods and results. This directly follows by characterizing the IFOG as an optical communications channel and the rotation as a channel (disturbance) estimation problem. None of this body of knowledge is exhibited in the prior literature up to and including Burns (1994). Of especial importance is the utility of employing the methodology of statistical communication theory for analyzing the random scattered light interference effects, the subject of the companion volume. The results of our analyses of the necessarily imperfect Serrodyne flyback effects are new (Appendix B). In contrast to the applied physics mindset dominating the research and results published to date, a communications engineering viewpoint runs throughout this report.

IFOG ARCHITECTURE

The IFOG, like its predecessor the ring laser gyro (RLG), depends on the Sagnac effect to sense a rotation rate. All Sagnac-effect-based rotation-rate sensors (including the RLG) establish a pair of contra-propagating light beams sharing a planar,[4] closed circuit, light guide having exquisite optical symmetry (i.e., "reciprocity") between the clockwise and counterclockwise propagation paths. Mechanical rotation of the plane containing the light circuit upsets the optical symmetry, thereby creating an observable signal. Photoelectronic processing of this signal is employed to accurately estimate the rotation rate.

---

[4]An RLG product (the ZLG or zero lock gyro) deliberately introduces a small geometric perturbation to the light circuit orthogonal to the normally planar path (i.e., almost planar) to quantum mechanically split the polarization modes of the propagating light beams. The desired result is to eliminate the need for mechanical dither.

Shown in Figure 1 is a representation of an IFOG with the following subsystems:

a. A rotation-sensing coil of single mode, *polarization preserving*,[5] optical fiber carefully wound on a coil form. Depending on the measurement accuracy desired, from 100 m to over 1 km of fiber is used.

b. A suitable, solid-state, wideband optical source, e.g., super luminescent diode.

c. An optical directional coupler and multifunction integrated optics chip (MIOC) to (1) divide and carry the optical source beam to the opposite two ends of the fiber coil where they will propagate as a clockwise (CW) and a counterclockwise (CCW) signal, (2) electro-optically modulate the light, (3) add together the exiting CW and CCW beams from the coil ends and carry the composite, vector-summed beam back toward the optical photodetector, and finally, (4) serve as a polarizer to select and orient the unpolarized source light onto the desired polarization axis of the polarization-maintaining fiber.

d. An electronics package to produce the desired electro-optic signal modulation and to implement the signal processing at the output of the photodetector.

Inserted between the optical source and MIOC (Figure 1) is an extremely high-quality (discrete component) optical directional coupler. This coupler must isolate the source beam entering the MIOC (on its way to the fiber coil) from contaminating the summed output beams (CW + CCW) exiting the MIOC (from the coil) en route to the photodetector. Ideally, this coupler would be incorporated on the MIOC. Unfortunately, MIOC technology cannot yet achieve the isolation levels needed between the optical source light and the returning CW + CCW composite beam.

---

[5]To date, navigation-grade IFOGs require a very high degree of optical reciprocity in the light path, necessitating operation in a single state of polarization on the desired axis of the fiber. Unfortunately, this increases the costs of the gyro; it does, however, simplify the mathematics!

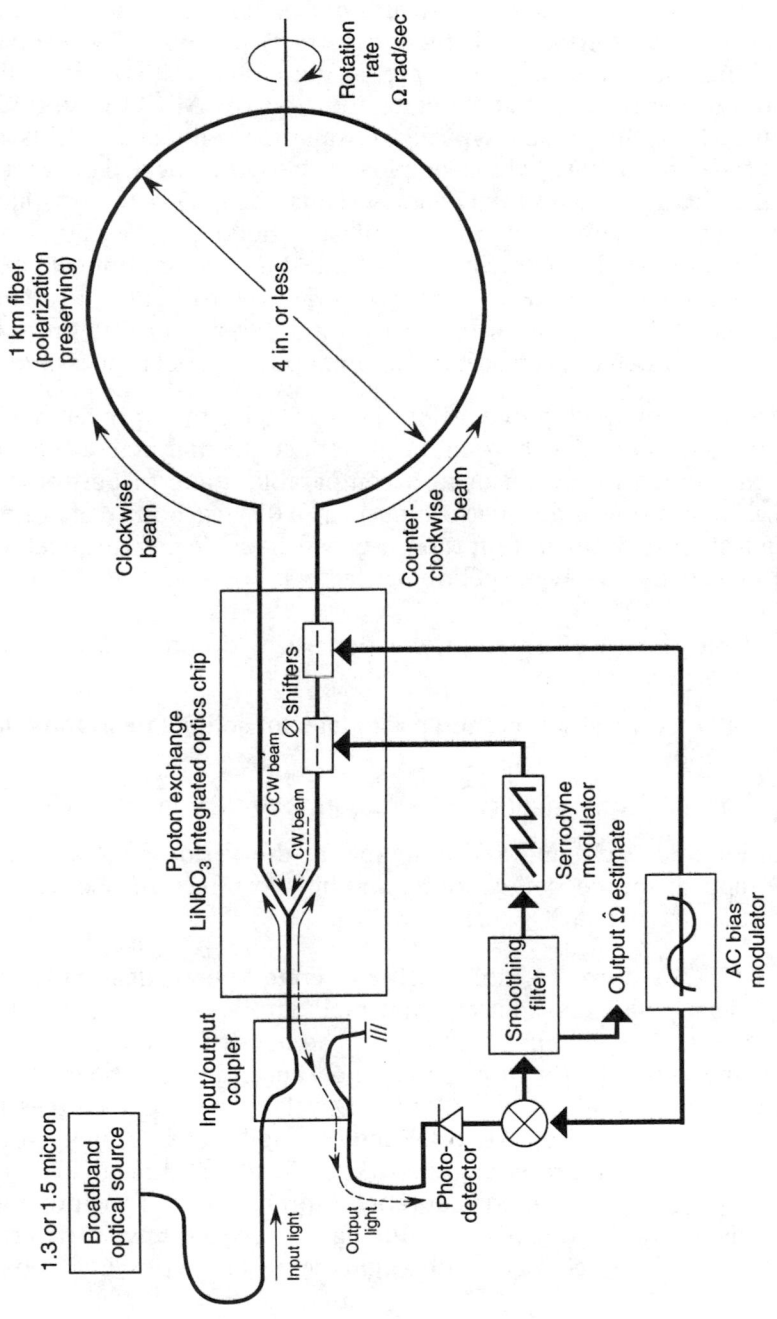

Figure 1—Closed-Loop IFOG Architecture

Since the light (i.e., photon stream) makes but one trip around the fiber rotation-sensing coil, the structure is not optically resonant. Lacking optical resonance, the propagation bandwidth of the IFOG is extraordinarily large, i.e., that of the fiber or MIOC waveguides, whichever is the smaller; typically, terrahertz in either case. This is in contrast to an RLG, which employs a closed optical path around which the light makes many passes. Thus, the RLG is a narrowband, optically resonant (ring cavity) optical structure. (Of lesser significance, the RLG lasing medium is within the resonant optical circuit whereas the IFOG optical source is exterior to the circuit.) Therefore, the mode of operation of the IFOG will differ significantly from the RLG (i.e., wideband versus narrowband optical signal processing).

The object of seeking to develop the IFOG, given the proven performance of the RLG, is to dramatically reduce the manufacturing costs of the rotation-rate sensor at a comparable grade of performance (e.g., navigation grade performance of 0.01 deg/hour drift rate or equivalently ~1 nm/hr drift rate). Success in achieving this goal rests on at least the following technology factors:

a. Optical parts purity and signal processing optimization to match RLG performance

b. MIOC, optical source, and photodetector solid-state foundry fabrication

c. Low-cost, high-quality, polarization-maintaining optical fiber

d. Robotic packaging and assembly of the various parts, such as optical source, coil winding, and fiber connections (source pigtail, pigtail to MIOC, coil to MIOC).

Early IFOGs were assembled using discrete fiber optical parts (e.g., couplers, polarizers, splitters, and modulators) operating with light sources at 0.83 micron wavelength. The net drift rate bias achieved was in the 0.1 deg/hr or more regime. Recently, brassboard IFOGs have been assembled using longer wavelength optical sources (1.3 micron or 1.5 micron), one kilometer length sensing coils wound with polarization preserving fiber (PPF), MIOC, and feedback architecture. The PPF coil considerably improves optical reciprocity of the light path by drastically reducing the level of unwanted cross-polarized (nonreciprocal path) light being crosscorrelated on the photodetector. Laboratory results have achieved bias drift less than 0.01 deg/hr. A lower grade of performance IFOG at even lower cost can be fabricated from the high-quality technology base by employing any or all of the following:

a. Shorter coil lengths (200 m instead of 1000 m)

b. Single mode fiber instead of polarization-preserving fiber

c. Less sophisticated signal processing

d. Open-loop instead of closed-loop architecture.

IFOG OPERATING PRINCIPLES

For a correct physical understanding of the Sagnac effect in both RLGs and FOGs, one must employ general relativity (Ezekiel and Arditty, 1982; Arditty and Lefevre, 1981; Chow et al., 1985). When rotating, the IFOG sensing coil is an accelerating frame of reference. Consequently, one cannot employ special relativistic reasoning. (It incorrectly predicts zero optical effect.) The discussion that follows is at the purely qualitative engineering level. A reader interested in the fundamental physical development may wish to consult Arditty and Lefevre (1981) and Chow et al. (1985).

The IFOG photodetector, through its rectification and short-term optical time-averaging characteristic, mechanizes the mathematical crosscorrelation[6] between the two contrapropagating optical output beams that are launched from the common optical light source and pass around the long fiber rotation-sensing coil on a single, desired axis of polarization. Ideally, with no optical path imperfections or asymmetries (i.e., with perfect optical reciprocity), the photodetector output measures the autocorrelation function of the light source (time delayed by passage through the fiber coil). With no rotation, the peak value of the optical source autocorrelation function is sensed. Imposing a mechanical rotation rate causes the propagation

---

[6]In general, the magnitude squared of the summed beams produces three terms: a DC term, a double optical frequency term, and an optical, time-averaged, videoband cross-product term. The DC term can be blocked, the double frequency term time averages to zero, and the cross-product video is the crosscorrelation of the two beams.

time of the contrapropagating beams in the sensing coil to become equally and oppositely displaced with respect to their at-rest delay. This produces an optical phase shift (as opposed to the doppler frequency shift in a RLG) between the two beams, and an off-peak value of the optical source autocorrelation function is then obtained at the photodetector output.

This simple, interferometric configuration produces three difficulties with the photodetector output signal. First, because all autocorrelation functions are even functions of delay, rotation direction cannot be directly sensed. Second, the rotation-rate sensor gain is low at low rotation input because the autocorrelation function (typically a raised cosine) has zero differential sensitivity at zero rotation input. Third, the crosscorrelation measurement at the output of the photodetector is centered around DC, requiring DC electronic amplification circuits. These difficulties can be corrected by introducing a frequency modulation at a multikilohertz frequency (e.g., 100 kHz) on each of the counterpropagating beams prior to photodetection as shown in Figure 1. This is accomplished with an optical phase modulator on one of the sensing coil leads. The drive voltage to the phase modulator periodically varies the optical path time delay through the modulator.[7] Note that one beam is phase-modulated before entering the fiber sensing coil and the other beam is modulated after exiting the coil. This produces optical phase modulations time-shifted (squinted) with respect to each other by the beam propagation time through the coil. This impressed optical phase modulation is referred to as an "AC bias" signal.

The photodetector crosscorrelation output can be processed around the first (or higher odd) harmonic of the AC bias modulation frequency by synchronously multiplying and smoothing the photodetector wideband video output signal by a replica of the AC bias signal (or higher odd harmonic) driving the optical phase modulator. The twofold consequence of this modulation/signal processing arrangement is to allow use of less-expensive AC photodetector output amplifiers centered on the imposed phase modulation subcarrier frequency and, more importantly, to produce a direction-

---

[7]Time base variation is mathematically equivalent to imposing a periodic phase or frequency modulation.

sensitive, high gain, skew symmetric detector output signal, which is zero for zero rotation input rate. (Rotation direction is given simply by the sign of this output.)

Although referred to as AC bias modulation by the IFOG community, the communication theoretic analogy is with frequency modulation (FM) discriminators. Recall that the AC bias is a phase modulation impressed as a subcarrier on the optical wave and that the sensing coil is an optical delay line. Since FM discriminators can be synthesized from delay lines, it will come as no surprise to see conventional FM results emerge from the model.

Now consider the long optical path length (on the order of 1 km) used by a high-sensitivity IFOG. Discrete imperfections in the optical path (for example, due to splices or connectors, nicks in the fiber cladding, or density variations and impurities in the fiber core) will cause small-scale optical reflections (backscatter), and crosspolarization mode coupling (forward scatter). These unwanted optical interference signals are analogous to the multipath interference in a radio link. In direct analogy with radio multipath, *spread spectrum optical sources* can be used to combat the multipath. Currently, all-solid-state examples of such optical sources include the super luminescent diode (SLD) and, possibly, a fiber light source derived from the erbium-doped fiber amplifier (EDFA) so popular in the telecommunications industry (Fesler et al., 1990, and Wysocki et al., 1991).

All such "optical multipath" interference sources separated from the desired optical signal or from each other by *less than* the coherence length of the source (i.e., the product of the optical source autocorrelation envelope decay time and the speed of light in the fiber core) will upset the measurement accuracy. This results from the photodetector crosscorrelating the exiting light beams (desired plus interfering) of the fiber coil. This crosscorrelation not only produces the desired autocorrelation function of the optical source but is also corrupted by crosscorrelations between the "multipath" generated beams and the desired beams. However, the crosscorrelation between sites separated *by more than the optical source's coherence length are essentially zero.* Short coherence length optical sources having large instantaneous bandwidth[8] (i.e., spread spectrum sources) considerably reduce the number of small interference signals corrupting the measurement accuracy. The IFOG photodetector output processing operates on the optical phase variation (raised cosine) of the optical source autocorrelation function and not on the slower varying envelope to measure rotation rate. Consequently, an optical source having large spectral bandwidth albeit narrowband compared to the optical frequency has a useful autocorrelation function for operating the IFOG.

As described above, the IFOG would suffer from a very limited measurement dynamic range. (That is to say, the ratio of the largest usefully measurable rotation rate to the random noise output will be small.) The cosinusoidal periodicity intrinsic to the autocorrelation phase function (the component of the autocorrelation function being processed) becomes 2π ambiguous at higher rotation rates. This effect produces measurement (not device) failure. Since the sinusoid is linear only for small values of its argument, the useful measurement dynamic range is even further restricted by the fact that the IFOG sensor gain (scale factor) becomes nonlinear with increasing input rotation rate. Increasing the desired measurement accuracy lowers the allowed rotation rate for linear scale factor, i.e., it further reduces the useful dynamic range.

The conventional method of increasing the dynamic range without losing accuracy is to employ measurement feedback. As the Sagnac-generated optical phase shift develops, it is sensed at the photodetector. Feedback electronics, at the photodetector output, filter the Sagnac signal and drive a voltage-controlled optical phase or frequency shifter (implanted on the integrated optics chip) to cancel out the developing Sagnac effect optical signal. Then the rotation-rate measurement can be taken from the drive input to the voltage-controlled optical shifter. Ideally, the optical shifter drive voltage is directly proportional to the rotation rate. Functionally, the Sagnac optical signal is used as the input error signal to an optical tracking feedback loop. In this regard, a feedback IFOG is analogous to the phase lock loop (PLL) so widely used in radio receivers.

---

[8]In theory, the source also could be broadbanded by electronically induced modulation. Alternatively, other signal processing techniques for multipath rejection might be employed.

Having outlined the general operating principles of a high-performance feedback IFOG, the next task is to develop a mathematical model for the requisite signal processing.

Chapter Two

DETERMINISTIC ANALYSIS

ASSUMPTIONS AND MODEL

In Figure 2, the optical structure of Figure 1 is represented by an equivalent pair of symmetric communications channels sharing a common optical delay line, i.e., the fiber coil of length L. The "transmitter" and "receiver" are collocated with the optical signals passing through the common optical delay line but in opposite directions. A mechanical rotation rate, Ω, upsets the channel optical symmetry. The problem is to design optical signal modulation and processing to utilize the deterministic disturbance to the channel symmetry induced by rotation rate to estimate Ω. Letting $\hat{\Omega}$ denote our (output) estimate of (input) Ω, the following properties are desired:

1. $\hat{\Omega}$ is unbiased ($E\hat{\Omega} = \Omega$).

2. $\hat{\Omega}$ is linear and sign-sensitive to Ω. Denote the electronics output variable by ξ. Then, in the ideal noise-free case, ξ equals $K\Omega$. The constant K is the scale factor, and $\hat{\Omega} \equiv K^{-1}\xi$ K must remain constant over environmentally changing conditions, e.g., temperature, acceleration, and rotation.

3. $\hat{\Omega}$ implementation should not require precision DC processing.

4. $\hat{\Omega}$ should not be sensitive to optical scatter processes (multipath).

The simple propagation model employed here first treats the fiber coil as an equivalent, ideal, lossless and scatter-free, dispersionless electromagnetic delay line. As mentioned previously, the analysis of

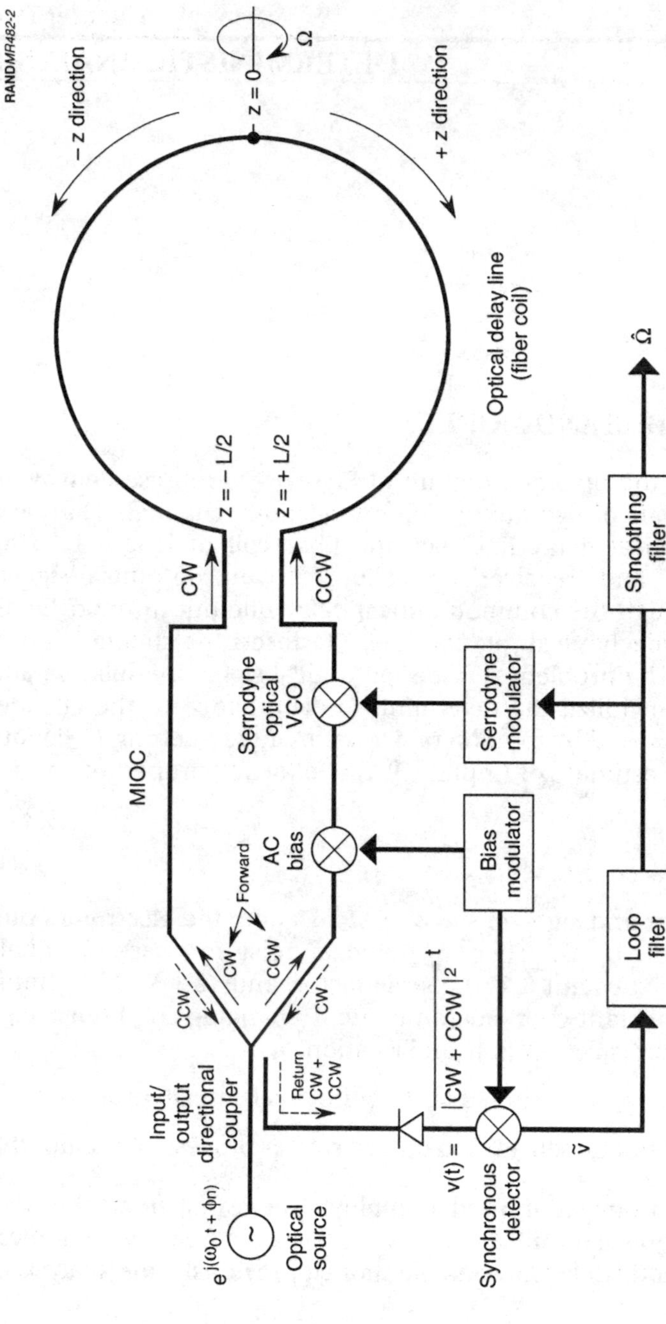

Figure 2—Equivalent Communications Channel

a coil mechanically rotating at rate Ω requires the use of the general theory of relativity (Ezekiel and Arditty, 1982; Arditty and Lefevre, 1981; Chow et al., 1995) because the light is propagating in an accelerating reference frame. If one expands the general relativistic solution in a power series in v/c and retains only the terms linear in v/c, one obtains (almost) the same result as if one treated the two counterrotating streams of Einsteinian photons as equivalent Newtonian corpuscles.

In the discussion that follows, the "corpuscular" mnemonic is employed to describe the propagation of the desired strong signal beams. In this volume, the weak optically scattered light terms are ignored to develop optical phase modulation useful in achieving an observable $\hat{\Omega}$ having the desired properties cited above. The companion volume treats the scattered light noise processes.

In the analysis for the signal beams, the phase content of the photodetector output depends on the difference between the physically developed (internal) Sagnac phase shift, ϕ_s, imparted by the rotation rate, Ω, and the (external) optical phase shift imposed by the closed-loop electronics driving an optical phase shifter embedded on the MIOC. The difference between these phase shifts is the error phase, ϕ_e, which is the variable driving the feedback loop. For a properly designed system in steady-state rotation at speed Ω the loop will drive ϕ_e to zero and the externally imposed phase shift will equal the Sagnac internal phase shift. Then Ω can be estimated by measuring the loop control variable.

NOISE-FREE SIGNALING MODEL

Referring to Figure 2, we have a "perfect" fiber optic delay line of physical length L wrapped onto a coil. To represent the propagation through the fiber, we choose the exact center of the fiber delay line as the reference point (z = 0) for the location z along the fiber. Arbitrarily, set the sign convention such that the CW beam enters the fiber delay line at z = –L/2 (early) while the CCW beam enters at z = +L/2 (late).

The optical source output is characterized, ideally, as a constant amplitude (fixed intensity) narrowband signal with a wideband noisy phase modulation process, ϕ_n. Thus, the source wave is given in complex notation by the following:

$$\text{source output} = \text{Re}\left\{\sqrt{2P_0}\exp j[\omega_0 t + \phi_n(t)]\right\} \qquad (1)$$

where

$$
\begin{aligned}
P_0 &= \text{optical source power} \\
\omega_0 = 2\pi c / \lambda_0 &= \text{the mean center frequency of the source and} \\
& \quad \lambda_0 \text{ is the mean wavelength} \\
\phi_n &= \text{wideband noise process} \\
\text{Re}\{\} &= \text{real part.}
\end{aligned}
$$

Now, let's follow the optical source output as it enters the IFOG coupler/MIOC/coil/photodetector assembly. The source wave proceeds unaffected through the coupler and enters the MIOC where its intensity (i.e., power) is halved and launched down the two parallel paths having "equal" path delays. (We assume that the two phase modulators on the lower path have a net zero average delay[1] through them.)

Before proceeding further, it is necessary to note that in all of the analysis in this volume, the two contrapropagating optical beams remain in the polarization state in which they are launched by the MIOC into the fiber optic rotation sensing coil. Consequently, it is not necessary to complicate the notation by indicating the polarization states for the deterministic signal analysis. In the companion volume, however, the scatter-generated "noise" light from the desired beams propagates around the coil in both states of polarization. Since cross-polarized forward scatter light is a major error source, it will be essential for the analysis in the companion volume to account for the polarization states. For the desired beams, the single state of polarization is maintained by the use of high-quality polarization maintaining (PM) optical fiber in the rotation sensing coil and by the exceptional polarization selection (polarizing) property of the proton exchange fabricated MIOC. Current MIOCs can provide upward of 60 dB extinction ratio; i.e., the ratio of light intensity on the unde-

---

[1]The net average delay incurred is physically balanced out by adjusting the optical length of the upper path.

sired state to that of the desired state at the output of the MIOC.[2] At the output of the MIOC, the two beams enter at opposite ends of the fiber coil.

The phase modulators on the MIOC change the index of refraction in the optical wave guides in proportion to the applied voltage by employing the electro-optic effect. The resulting time delay variation in the beam passing through the phase modulator is proportional to the externally applied voltage. This in turn produces an equivalent phase modulation on the beam equal to the product of the net delay induced by the modulator and the optical frequency ω_0. Thus, the two signal beams, CW and CCW, each entering at its coil input end can be represented by .

$$CW = \sqrt{P_0} \exp j[\omega_0 t + \phi_n(t)]$$

$$CCW = \sqrt{P_0} \exp j[\omega_0 t + \phi_n(t) + \phi_m(t)]$$

where $\phi_m(t)$ = externally applied optical phase modulation.

The phase modulation ϕ_m will be the linear sum of two separate processes: (1) a fixed externally induced modulation, independent of rotation rate, Ω, used to generate a direction-sensitive (to the sign of Ω) output with "AC" photodetector electronics (i.e., "AC bias") and (2) a modulation proportional to Ω used to cancel the Sagnac phase shift (i.e., to close the optical feedback loop). Additionally, the AC bias will produce rotation-rate sensor gain at low mechanical rotation input but at some loss of detected light.

When the coil is at rest ($\Omega = 0$), the two optical paths are identical and both beams will take the same time, $\delta_0 = Ln/c$ seconds, to transit the coil. Here, c is the speed of light in a vacuum and n is the index of refraction of the fiber core. Again referring to Figure 2, the CCW wave *enters* the coil already modulated by ϕ_m whereas the CW wave

---

[2]More precisely, for polarized light at the input to the MIOC, the ratio of undesired to desired polarization at the MIOC output is the product of the MIOC extinction ratio and the degree of polarization (including its alignment in the MIOC input "pigtail") in the input light.

exits the coil and then is modulated by ϕ_m *but at δ_o seconds later.* If we redefine the origin of our time axis to the input of the photo-detector, $t \leftarrow t + \delta_0$, we can represent the two beams by

$$CW = \sqrt{P_0}\, exp\, j[\omega_0 t + \phi_n(t) + \phi_m(t - \delta_0)]$$

$$CCW = \sqrt{P_0}\, exp\, j[\omega_0 t + \phi_n(t) + \phi_m(t)] \tag{2}$$

The key point here is the δ_0 sec of "squint" between the contrapropagating beams in the externally induced AC bias phase modulation ϕ_m.

Next, we classically model the photodetector output as the short time average of the intensity of the composite beam impinging on the photodetector. Mathematically, the photodetector output, $v(t)$, is given by

$$v(t) = \frac{1}{2} \overline{|\,CW + CCW\,|^2}^{\,\omega_0 t} \equiv \frac{1}{2} \overline{|\,CW\,|^2}^{\,\omega_0 t} + \frac{1}{2} \overline{|\,CCW\,|^2}^{\,\omega_0 t} \tag{3}$$
$$+ Re\left\{ \overline{CCW * \cdot CW\, |}^{\,\omega_0 t} \right\}$$

where CW and CCW are defined by Eq. (2), the asterisk denotes complex conjugation and the overbar indicates short-term time average over several hundred optical cycles of $\omega_0 t$. It is assumed here that the desired beams have ideal constant envelopes containing no intensity fluctuations. Then,

$$v(t) = P_0 + Re\left\{ \overline{CCW * \cdot CW\, |}^{\,\omega_0 t} \right\}$$

From Figure 2, the combination of the synchronous detector and loop filters serves to remove the non-information-bearing DC term, P_0, from the detector output, $v(t)$, leaving only the complex conjugate cross-correlation term between the CW and CCW beams. This leads to the desired characterization of the photodetector as a cross-correlator, i.e.,

$$v(t) = Re \; \overline{CCW * \cdot CW}^{\omega_o t} \tag{4}$$

(However, when there are noise and/or intensity fluctuations in the composite beam impinging on the photodetector, one must return to Eq. (3) as the $| CW |^2$ and $| CCW |^2$ terms need not be time constant.)

Inserting Eq. (2) into Eq. (4) produces

$$v(t) = P_o \cos[\phi_m(t) - \phi_m(t - \delta_o)] \tag{5}$$

Note the complete disappearance in Eq. (5) of not just the optical frequency term, $\omega_o t$, but *especially* the optical source phase noise term, $\phi_n(t)$. This highly desired result is a direct result of the assumed ideal mathematical symmetry (optical reciprocity) of the CW and CCW propagation paths. From the point of view of communication theory, we are communicating with ourselves through the long fiber coil and cross-correlating a received spread spectrum signal (from one end of the coil) with a duplicate reference spread spectrum signal from the other end of the coil. In other words, we are executing a "conventional" despreading operation[3] with our pair of optical signals.

In most conventional optical channels (i.e., without a collocated transmitter receiver pair), the phase noise, ϕ_n, in the optical transmitter is highly deleterious. Since ϕ_n cannot be replicated or stored at the distant receiver for despreading, it can serve no purpose other than to degrade the noise margin for the channel. However, for the IFOG with a collocated "transmitter/receiver," the ϕ_n process is used to significantly mitigate the optical scatter noise generated by imperfections in the propagation path. This will be the main subject of the companion volume. Suffice it to say here that spread spectrum sig-

---

[3]The sophisticated reader will note that the time delay modulation executed by the phase modulator(s) also slightly time-offsets the optical source phase noise process $\phi_n(t)$ and conceivably could lead to a loss in the "spectrum despreading." In fact, this loss is negligible. From Appendix A, the $\Delta \phi_n(t)$ process between the CW and CCW waves inserted by the external phase modulator $\phi_n(t) - \phi_n(t - \delta_o)$ has peak time difference $\Delta(t) - \Delta(t - \delta_o)$ on the order of $1.8/2\pi \; \delta_o$ seconds. That is to say, the peak time difference is on the order of one wave length of optical RF. Even though ϕ_n is a very broadband process, it is still very narrowband compared to ω_o (or $\omega_o/1.8$). Thus, the peak time shift is negligible and $\Delta \phi_n \cong 0$.

nals are conventionally used in radio communication systems to reject multipath interference and the scatter noise is a form of multipath interference.

We will employ an engineering "mnemonic" (as opposed to a correct
general relativistic derivation; for example, see Ezekiel and Arditty
(1982), Burns (1994), Arditty and Lefevre (1981), and Chow et al.
(1985)) to write out the Sagnac phase shift ϕ_s developed by the fiber
optic sensing coil rotating at Ω rad/sec. First, we will write out ϕ_s as
if the photons were Newtonian corpuscles. To first order, this
produces an almost correct expression for ϕ_s. We then make the
correction by which the relativistic effect is manifested.

For a fiber sensing coil of effective diameter, D, with a length of fiber,
L, rotating at a fixed rate, Ω rad/sec, the coil ends will move a dis-
tance, Δ, equal to $(D/2) \Omega \delta_0$ in the time, δ_0, that it takes a beam
to traverse a stationary coil (i.e., when $\Omega \equiv 0$). Employing the
"corpuscular" mnemonic, the Newtonian light corpuscles traveling
in the beam moving in the direction of the mechanical rotation must
travel a path Δ meters longer and the light corpuscles in the beam
moving against the rotation must travel a path Δ meters shorter. If
we let n denote the index of refraction in the fiber core (on the desired polarization axis), then the average speed of the desired signal
light is c/n, where c is the speed of light in vacuum. If we divide Δ by
the average speed, c/n, of the corpuscles in the fiber core, we obtain
the time difference between the two contrapropagating beams
caused by the Ω rotation. Multiplying by the optical frequency, ω_0,
gives the accumulated Sagnac phase shift induced on the output
beams by the rotation rate Ω. Notational tradition denotes the
Sagnac phase shift, ϕ_s, as the total phase difference between the
CCW and CW beams, which is evenly split between the two beams.
Then the (almost correct) phase difference in each beam, $\phi_s /2$, is
given by

$$\phi_s /2 = (D/2)\Omega \delta_0 \omega_0 n / c$$

Since $\omega_0 \equiv 2\pi c / \lambda$, $\omega_0 \delta_0 = 2\pi n L / \lambda$, and we have (the almost
correct expression)

$$\phi_s = \left[2\pi \frac{n^2 LD}{\lambda c} \right] \Omega$$

where λ is the mean wavelength of the source light in a vacuum.

That this version of ϕ_s needs correcting is mandated by recalling that the light beams are traveling through an angularly accelerating medium, i.e., the fiber core. A simplistic engineering derivation of the correction is given in Burns (1994, Chap. 1, pp. 1–30), where the effective speed of light in the fiber core is adjusted for the rotation rate, Ω, *and* this adjustment depends on whether the beam is moving with or against the motion of the fiber core. A careful derivation is presented in Arditty and Lefevre (1981) (and repeated in Ezekiel and Arditty (1982)). The net effect is that the expansion for the equivalent speed of light has the form

$$c[1 + \sum_k \text{small coefficient} \cdot n^{-2k}]$$

The odd orders in n are canceled by the beams' sensitivities to the direction of the rotating fiber core. So to first order, the equivalent index of refraction to be used in calculating ϕ_s is $n \equiv 1$.

Thus, the relativistically corrected ϕ_s is well approximated by

$$\phi_s = \left[2\pi \frac{LD}{\lambda c} \right] \Omega \tag{6}$$

From Eq. (6) the total Sagnac phase shift is proportional to Ω (and its sign) with "scale factor" (proportionality constant) given by the bracketed term. The job of the optical signal processing is to extract ϕ_s in as low noise, zero bias, and linear a manner as possible. While extracting ϕ_s we must keep LD/λ fixed (especially λ, the mean wavelength of the optical source); otherwise, we will be dividing the "linearly extracted" ϕ_s by an incorrectly calibrated scale factor and will produce an incorrect estimate of Ω.

AC BIAS MODULATION AND OPEN-LOOP MODEL

Again, referring to Figure 2, we arbitrarily assign clockwise rotation to be $\phi_s > 0$, which is consistent with increasing z. Then Eqs. (2) and (5) are modified by including $\phi_s / 2$ in the following manner:

$$CW = \sqrt{P_o}\, exp\; j[\omega_o t + \phi_n(t) + \phi_m(t - \delta_o) - \phi_s / 2]$$

$$CCW = \sqrt{P_o}\, exp\; j[\omega_o t + \phi_n(t) + \phi_m(t) + \phi_s / 2] \qquad (2')$$

and

$$v(t) = P_o \cos[\phi_m(t) - \phi_m(t - \delta_o) + \phi_s] \qquad (5')$$

$$\phi_s = \left[2\pi \frac{LD}{\lambda c} \right] \Omega$$

Now, if $\phi_m(t) \equiv 0$, then v(t) is a DC term that will be rejected by our DC block. Alternatively, if we used DC electronics, v(t) = $P_s(1 + \cos \phi_s)$ is an even function in ϕ_s with zero (slope) gain at $\Omega = \phi_s = 0$. Thus, we should employ a bias modulation to obtain an odd function in ϕ_s with output gain at $\Omega = 0$.

Note that whatever ϕ_m that may be imposed externally, the coil physical architecture produces a finite difference phase process at the photodetector, $\phi_m(t) - \phi_m(t - \delta_o)$. Dividing this phase difference by δ_o estimates the derivative of ϕ_m and therefore approximates the frequency modulation that produces ϕ_m. Initially, we consider the IFOG operating in an open-loop mode so that $\phi_m(t)$ contains only the AC bias modulation. In Appendix A, a sinusoidal AC bias modulation is analyzed: $\phi_m(t) = (-A / 2) \sin \omega_m t$. Other, more advanced, forms of AC bias modulation are possible. From Appendix A, the synchronous detector output, v, is shown to be the following

$$\tilde{v} = P_o J_1(A) \sin \phi_s \qquad (7)$$

where $J_1(\bullet)$ is the Bessel function of order 1 and ω_m is chosen such that $\omega_m \delta_o$ = odd integer $\bullet \pi$. The output, v, is maximized at $A \equiv 1.8$ and

$$J_1(1.8) \equiv 0.5815$$

The sinusoidal AC bias modulation produces a DC-free direction-sensitive output with a gain slope (at $\Omega = 0$) given by (0.58) $P_0[2\pi LD / \lambda c] \equiv K_{SF}$. However, almost half the available signal light (42 percent) is lost in accomplishing this goal.

So long as ϕ_s, as given by Eq. (7), is small, $\sin \phi_s \cong \phi_s$. If we multiply $\tilde{v}$ by the scale factor $\hat{K}_{SF}$, given by Eq. (A.6) in Appendix A, we obtain our estimate, $\hat{\Omega}$. Unfortunately, this simple, open-loop structure breaks down if ϕ_s gets large (large Ω). Not only does $\sin \phi_s$ become nonlinear, it becomes 2π ambiguous. A solution to this problem is to operate the IFOG in a closed-loop manner producing an optical phase locked loop.[4]

FEEDBACK IFOG WITH SERRODYNE MODULATION

For a closed-loop (feedback) IFOG, an additional phase modulation on the MIOC must be introduced to null out the Sagnac phase shift, ϕ_s, developed by the mechanical rotation rate, Ω. Optimally, the nulling signal will be obtained by first estimating ϕ_s, as in the previous subsection, and then using this estimate to cancel ϕ_s. The resulting estimate, $\hat{\Omega}$, then should be proportional to Ω over a much larger range than is $\sin \phi_s$. The estimate of the rotation-induced Sagnac phase shift, ϕ_s, need not always be exact. Define the error phase, ϕ_ε, as the difference between the rotation-induced Sagnac phase shift, ϕ_s, and the "voltage controlled" external phase shift obtained from estimating ϕ_s and introduced by the (external) loop closure portion of the IFOG. In simple terms, ϕ_ε replaces ϕ_s in Eq. (7). As long as the "loop is in lock," ϕ_ε will be small enough that $\sin \phi_\varepsilon \approx \phi_\varepsilon$ even when Ω (hence ϕ_s) is large. The object of the phase lock loop (PLL) is to keep ϕ_ε dynamically small and drive it to zero when Ω is constant.

Invariably, all PLL implementations require a voltage-controlled oscillator, VCO. In our case the VCO output must be an optical wave

---

[4]The reader is assumed to have a casual familiarity with the basic operating principles of the PLL. The pedagogy for the feedback IFOG proceeds by analogy to the phase lock loop. For example, see Viterbi (1966).

whose frequency (or wavelength) is adjusted by the DC control voltage. Note that a frequency shift is equivalent to adding an unbounded phase ramp to the argument of the VCO output optical wave. The slope of this ramp is equal to the frequency shift, i.e., the slope of the phase ramp is controlled by the DC input voltage.

Unfortunately, it is *not possible* to physically execute this VCO function employing the electro-optic effect in a MIOC. (In fact, one would need an off-chip, bulk-optic frequency shifter, e.g., an acousti-optic Bragg cell.) Since the electro-optic effect on the MIOC can achieve only a bounded time delay,[5] the phase ramp must be periodically reset, producing a serrated (or serrodyne) waveform, as shown in Figure 3a. The serrodyne phase modulation mechanizing the VCO function on the MIOC is shown in Figure 3b. It is added directly to the AC bias modulation.

Continuing the analogy with FM theory, the output phase of any phase modulator can be mathematically represented by the indefinite integral of a time-varying frequency, $\eta(\tau)$.

$$\phi(t) = -\int^t \eta(\tau)d\tau \tag{8}$$

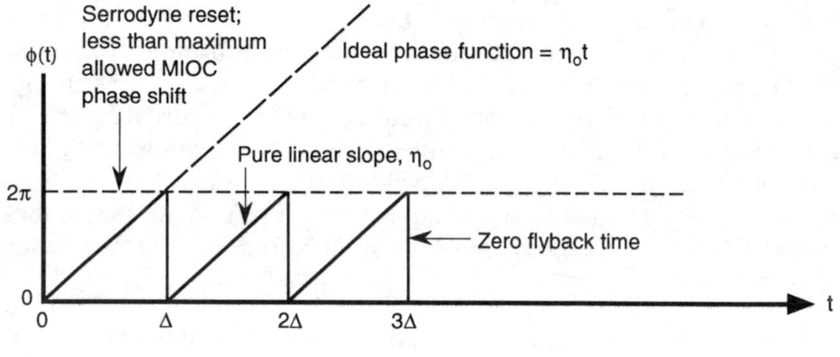

Figure 3a—Ideal Serrodyne Waveform

---

[5]Given by the product of the peak change of index of refraction and the optical length of the phase shifter.

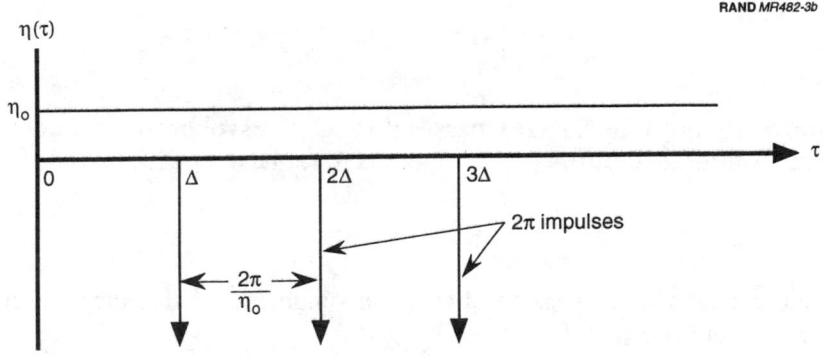

RAND *MR482-3b*

Figure 3b—Frequency Function, $\eta(\tau)$

In our case, $\eta(\tau)$ is the optical frequency modulation waveform outputting from the serrodyne modulator and is given by the derivative of the serrodyne phase waveform (Figure 3a). Shown in Figure 3b, $\eta(\tau)$ is seen to be in the superposition of a DC value η_0 and an impulse train caused by the step flyback resets in $\phi(t)$. With a slowly changing control voltage, η_0 is proportional to the control voltage, as is the impulse repetition frequency. The area under the impulse is equal to the value of phase growth chosen (i.e., 2π) at which to trigger the flyback reset. Now add the serrodyne modulation to the periodic[6] AC bias modulation to produce an aggregate phase modulation, $\phi_m(t)$, given by

$$\phi_m(t) = -\left[\int^t \eta(\tau)d\tau + (A/2)\sin\omega_m t\right] \tag{9}$$

Inserting Eq. (9) into Eq. (2') then produces Eq. (5) with ϕ_m now given by Eq. (9) above. Moreover, the phase difference, $\phi_m(t) - \phi_m(t - \delta_0)$, then fixes the lower bound on the "floating" phase integral of Eq. (9) and we obtain[7]

---

[6]Note that the AC bias peak phase must be bounded and the sum of the bias and serrodyne phase peaks must be less than the limiting value achievable on the MIOC.

[7]The astute reader will notice that the frequency integral (i.e., phase) in the argument of the cosine function in Eqs. (10) and (11) is merely the linear impulse response of a

$$v(t) = P_o \cos\left\{\left[\phi_s - \int_{t-\delta_o}^t \eta(\tau)d\tau\right] - A\sin\omega_m t\right\} \qquad (10)$$

Now $v(t)$, given by Eq. (10), passes through the synchronous detector and is filtered about ω_m. In Appendix A, replace ϕ_s with

$$[\phi_s - \int n(\tau)d\tau]$$

and repeat the process to obtain the synchronous detector output $\tilde{v}(t)$ given by Eq. (11).

$$\tilde{v} = P_o J_1(A)\sin\left[\phi_s - \int_{t-\delta_o}^t \eta(\tau)d\tau\right] \qquad (11)$$

Consider the ideal (unfortunately MIOC nonrealizable) case of an optical VCO (Voltage Controlled Oscillator) needed to implement a phase lock loop. Let the VCO output frequency be held constant so that $\eta(t) \equiv \eta_o$. Then

$$\int_{t-\delta_o}^t \eta(\tau)d\tau = \eta_o\delta_o$$

and

$$\tilde{v} = P_o J_1(A)\sin\phi_\varepsilon \qquad (12)$$

with

$$\phi_\varepsilon \triangleq \left[\phi_s - \int_{t-\delta_o}^t \eta(\tau)d\tau\right]$$

Thus, when δ_o is set equal to ϕ_s/δ_o, ϕ_ε goes to zero! This is exactly the condition for a PLL to stabilize. Consequently, for an ideal VCO one can introduce the mathematical methods of PLL feedback theory as presented, for example, in Viterbi (1966).

---

delay line of (time) length δ_o, which to first order is exactly what the fiber sense coil is. The Ω rotation places a skew symmetric perturbation on the δ_o delay.

Since the MIOC physically has a peak phase shift, implementing an ideal VCO is not possible. Consider a serrodyne waveform as depicted in Figure 3a with slope η_0 with flyback times at $t = \ell\Delta$; $\ell = 0$, $\pm 1, \pm 2, \ldots$. We are going to hold constant the peak ramp phase value, $\eta_0\Delta$, of the serrodyne waveform equal to 2π. As Ω slowly changes, so too will the value of η_0 needed to null ϕ_s change. Since we are holding $\eta_0\Delta \equiv 2\pi$, Δ will change as $2\pi/\eta_0$.

The angular frequency function, $\eta(\tau)$, for this serrodyne phase waveform is given by

$$\eta(\tau) = \eta_0 - \sum_{\ell=-\infty}^{\infty} 2\pi\delta(\tau - \ell\Delta) \tag{13}$$

$$\Delta = 2\pi/\eta_0$$

$$\delta(\bullet) \equiv \text{Dirac impulse function}$$

Figure 3b depicts $\eta(\tau)$ as the desired Ω dependent, constant frequency value η_0 plus an unwanted impulse train. The output $\tilde{v}$ in Eq. (11) requires the short-term δ_0 time integral of Eq. (13), given by

$$\int_{t-\delta_0}^{t} \eta(\tau)d\tau = \eta_0\delta_0 - 2\pi \sum_{\ell=-\infty}^{\infty} \int_{t-\delta_0}^{t} \delta(\tau - \ell\Delta)d\tau$$

$$= \eta_0\delta_0 - 2\pi \sum_{\ell=-\infty}^{+\infty} \text{rect}_{\delta_0}(t - \ell\Delta) \tag{14}$$

where the rectangular unit pulse function is defined as

$$\text{rect}_{\delta_0}(t - \ell\Delta) = 1; \text{ for } 0 < t - \ell\Delta \leq \delta_0$$

$$= 0; \text{ otherwise}$$

Thus, the desirable part of the approximating serrodyne modulation in Eq. (14) is the leading $\eta_0\,\delta_0$ term; the penalty for using the serrodyne is the impulse pulse train of period Δ with the area under each pulse equal to 2π. These impulses are the direct consequence of the serrodyne flyback needed to keep the peak serrodyne phase fixed to the value 2π. Unfortunately, the impulse train, with an area of 2π, is far from negligible. We will need to seek ways to mitigate the disturbances introduced to IFOG operation by the serrodyne flyback.

If we now again repeat the argument in Appendix A, but use Eq. (14) for the phase, we get for the output of the synchronous detector v, as follows:

$$\tilde{v} = P_0 J_1(A) \sin \phi_\varepsilon \qquad (15)$$

and

$$\phi_\varepsilon = [\phi_s - \eta_0 \delta_0] + 2\pi \sum_{\ell=-\infty}^{\infty} \text{rect}_{\delta_0}(t - \ell\Delta)$$

As in the ideal case, we still obtain the term $[\phi_s - \eta_0 \delta_0]$, which is driven to zero by setting $\eta_0 \rightarrow \phi_s / \delta_0$. However, there is added the unwanted pulse train; but it instantaneously step switches between the values 0 and 2π radians. If we carefully control the optical parameters to keep the peak serrodyne phase always at 2π radians and achieve an instantaneous flyback, the sinusoid in Eq. (15) is transparent to the 2π radian step changes in its argument. In this ideal case, Eq. (15) reduces to the desired Eq. (12).

Now the closed-loop operation drives $\eta_0 \delta_0 = \phi_s$. Since $\eta_0 \Delta$ is fixed to equal 2π, the flyback frequency Δ^{-1} equals $\phi_s / 2\pi\delta_0$. Inserting Eq. (6) for ϕ_s relates the flyback frequency to the rotation rate Ω through the following scale factor.

$$\Delta^{-1} = \left[\frac{LD}{c\lambda} \right] \Omega \qquad (16)$$

As an alternative to measuring η_0, one can measure the flyback frequency, Δ^{-1} (via simple digital counting circuit), and employ the scale factor exhibited by Eq. (16).

The success of the serrodyne modulation depends on several factors, including the following:

1. Stable and accurate value of LD/ $\lambda\delta_o$, especially the λ value

2. Stable and accurate trigger threshold, $\eta_o\Delta \equiv 2\pi$

3. A very prompt serrodyne flyback time (ideally instantaneous).

Theoretical study of serrodyne impairments must include item 3 above, which can introduce distortion terms.

Chapter Three

SERRODYNE-INDUCED IMPAIRMENTS

METHOD

In this chapter we provide a qualitative synopsis of Appendix B, which analyzes the effects produced by a real, noninstantaneous, serrodyne flyback waveform. Proceeding from the idealized, instantaneous flyback serrodyne, the equivalent FM representation of the serrodyne flyback-generated impulse train becomes a periodic train of real pulses. Each pulse has small, nonzero time duration but, as in the ideal case, still contains an "area-value" of 2π. Employing the lossless optical delay line model for the fiber rotation sensing coil allows one to characterize the effect of the coil on the serrodyne flyback waveform as a "moving window" time average of the actual flyback pulse train where the window width is the one-way propagation time, δ_0, through the fiber coil.

If the flyback is perfect, i.e., impulse of 2π, the moving window average produces for each impulse a perfect rectangle of width δ_0 and height 2π. Then, the IFOG photodetector output from the periodic serrodyne flyback switches exactly (and instantaneously) by 2π values, i.e., the flyback has no effect. We can thus take our real flyback pulse output from the coil and best fit it with a perfect 2π rectangle of width δ_0. The difference waveform, e(t), is that component of the real flyback that will cause distortion at the output of the synchronous detector.

The flyback error waveform, e(t), will generally be a pair of periodically occurring antisymmetric spikes or dipole-pulse; one minus spike followed by a plus spike separated from each other by δ_0 and each lasting less than δ_0 sec. This pattern repeats at the serrodyne flyback period Δ, which is inversely proportional to the sensed mechanical rotation rate, Ω. When $\eta_0 \delta_0 = \phi_s$, the output from the photodetector is then given by the following:

$$v(t) = P_0 \, \text{Re}\left\{ e^{jA \sin \omega_m t} e^{-j2\pi \sum\limits_{\ell=-\infty}^{+\infty} e(t-\ell\Delta)} \right\} \tag{17}$$

where

$$\sum_{\ell=-\infty}^{+\infty} e(t - \ell\Delta) = \text{the periodic dipole error waveform.}$$

Now replace the time representation of the error waveform with its Fourier series representation and observe that the exponential of an infinite sum is the infinite product of exponentials, to rewrite Eq. (17) as

$$v(t) = P_0 \, \text{Re}\left\{ e^{jA \sin \omega_m t} \prod_{n=1}^{\infty} e^{-ja_n \sin n\theta} \right\}$$

where the a_n are the Fourier coefficients of the periodic error waveform, $e(t)$, and θ equals $(2\pi / \Delta)t$. There are then two paths open. The simplest, for first-order results, employs a term-by-term Taylor series expansion of $e^{-ja_n \sin n\theta}$. The second, for third-order results, uses the Fourier Angier Bessel series expansion for each $e^{-ja_n \sin n\theta}$. Appendix B pursues both paths and "passes" the results through the synchronous detector to determine the output distortion effects.

FIRST-ORDER RESULTS

The harmonic amplitudes or Fourier coefficients, a_n, of the error waveform must be kept small. The higher harmonic amplitudes decrease as $1/n$. The lower ones must be minimized by keeping the duration of each spike as short as possible. Using a Taylor series expansion for each "harmonic" exponential in the infinite series and retaining only first- and second-order terms from the product yields an approximate infinite product with the following form:

$$\prod \cong 1 - j \sum_{n=1}^{\infty} a_n \sin n\theta + \text{real second harmonic terms}$$

$$= 1 - j \sum_{\ell=-\infty}^{\infty} e(t - \ell\Delta) + \text{real second harmonic terms}$$

(18)

The effects produced on the output of the synchronous detector are summarized below.

1. Letting $\overline{d}$ be 1/2 the mean square power in e(t), the net output is reduced to $[1 - \overline{d}]$. Hence, the serrodyne loop gain is reduced to $1 - \overline{d}$. Therefore, $\overline{d}$ must be kept small compared to 1.

2. The second harmonic terms in $\prod$ are purely real. Consequently, they are in quadrature to the synchronous detector, which is sensitive only to the imaginary components. Therefore, not only are these pure real terms double frequency but they are blocked by the synchronous detection process.

3. Unfortunately, the periodic error waveform comes riding right through the synchronous detector. It must be kept small, which in turn will guarantee keeping $\overline{d}$ small. This is the dominant effect of the serrodyne flyback!

THIRD-ORDER RESULTS

Expanding each harmonic exponential in the infinite product in a Fourier Angier Bessel series and executing the approximations by retaining terms to only third order produces the following results:

1. The first-order results above are reproduced directly.

2. There are third-order additional corrections to the first- and second-order terms, which are at least $5 \overline{d}$ or more smaller and thus negligible.

3. There is an additional small imaginary third-order term to which the synchronous detector will be sensitive but it should be

insignificant in comparison to the first-order distortion. (See Appendix B, Eqs. (B.18) and (B.18'.)

The dominant impairment is the direct, essentially linear, leak-through of the error waveform, $e(t)$, by the synchronous detector. Consequently, the design of the serrodyne IFOG must keep the fly-back as sharp (wideband) as possible.

Chapter Four

CONCLUSION

In this volume, we have shown that employing the classical propagation model with strong optical signals allows one to bring into play well-established communication theoretic methodology for analyzing a feedback IFOG. We obtained an analogous frequency modulation (FM) representation for the effect of the electro-optic phase modulator, which then directly leads to time-honored FM analysis for optimizing the AC bias modulation. This directly leads to optimally setting the AC bias modulation frequency and amplitude values.

Further, we were able to show that feedback operation of the IFOG can employ well-known phase lock loop analysis with the exception that the theory must be modified to account for the limitations of the realizable optical serrodyne voltage-controlled oscillator on current integrated optic chip technology. We then gave a simplified analysis for the serrodyne-induced distortion terms. The principal result is the unattenuated presence of the serrodyne flyback error waveform at the output of the synchronous detector.

In mathematically describing the IFOG, we also set up the "spread spectrum" nature of the optical source and the crosscorrelating "despreading" operation of the photodetector. By now the reader will have noticed that we have yet to exploit analytically the benefits usually attendant to the use of spread spectrum signals in communication channels, i.e., interference and multipath rejection. Indeed, these benefits also powerfully come into play for the IFOG by rejecting the optical light scattered by the desired signal induced throughout the IFOG optical paths. These are random-noise-like interference signals and their analysis will be presented in the companion volume.

FM ANALYSIS OF BIAS MODULATION

Here we analyze a sinusoidal AC bias modulation. Other more advanced forms of bias modulation are possible. Because of the electro-optic effect in the integrated optic chip, applying a positive voltage, b, to the on-chip phase modulator increases the time delay through the modulator by a factor proportional to b, and thereby produces an equivalent phase shift proportional to $-\omega_o b$. By appropriate choice of a sinusoidal drive voltage, b, we can achieve a phase modulation, $\phi_m(t)$, given by

$$\phi_m(t) = -(A / 2) \sin \omega_m t \qquad (A.1)$$

where the amplitude, A, and frequency, ω_m, of the impressed AC bias modulation will be chosen to maximize performance as shown below. Inserting $\phi_m(t)$ into Eq. (5) of the text produces

$$v(t) = P_o \cos[\phi_s - (A / 2)(\sin \omega_m t - \sin \omega_m(t - \delta_o))] \qquad (A.2)$$

First, we determine the optimum bias frequency. From AC it is clear that if we choose $\omega_m \delta_o$ to be equal to an odd integer of π or f_m = (odd integer) $/2\, \delta_o$, then v(t) simplifies to

$$v(t) = P_o \cos[\phi_s - A \sin \omega_m t] \qquad (A.3)$$

By this choice of ω_m (in terms only of the coil transit time, δ_o, and independent of Ω) the complicating quadrature output components of the photodetector are eliminated.

Having so easily determined ω_m, we will need to work a little to extract ϕ_s and then seek a criterion to optimize the value of A. Since Eq. (A.3) resembles an FM output signal, employ the Jacobi-Angier Fourier Bessel series (Abramowitz aned Stegun, 1964) to obtain the following expansion on odd and even harmonics of ω_m.

$$v(t) = 2P_0 (\sin \phi_s) \left\{ \sum_{k=0}^{\infty} J_{2k+1}(A) \sin[(2k + 1)\omega_m t] \right\}$$

$$+2P_0 (\cos \phi_s) \left\{ \frac{1}{2} J_0(A) + \sum_{k=1}^{\infty} J_{2k}(A) \cos[2k\omega_m t] \right\}$$

(A.4)

where $J_K(\cdot)$ is the Bessel function of order k.

Observe that the Ω direction-sensitive term, $\sin \phi_s$, can be extracted from only an odd harmonic of the induced optical modulation.[1] Now, to extract the $\sin \phi_s$ term, the first harmonic, $k = 0$, will produce the largest amplitude coefficient, J_1. Synchronous detection (i.e., multiplying the photodetector output, v(t), by $\sin \omega_m t$ and time-smoothing or low-pass-filtering) of the product yields

$$\tilde{v} = P_0 J_1(A) \sin \phi_s$$

(A.5)

Equation (A.5) tells us to maximize the $\tilde{v}$ output on $\sin \phi_s$ by maximizing the value of $J_1(A)$, which occurs for $A \equiv 1.8$ and

$$J_1(1.8) \equiv 0.5815$$

This simple modulation scheme produces a direction-sensitive output with a gain slope (at $\Omega = 0$) given by

$$K_{SF} \equiv (0.58)P_0 [2\pi LD / \lambda^c]$$

(A.6)

---

[1]The $\cos \phi_s$ term can be extracted from an even harmonic; this can be of interest if one wishes to estimate P_0, the gyro output optical power impinging on the photodiode; for when ϕ_s is small, $\cos \phi_s \equiv 1$.

However, it does cost almost half the available signal light (42 percent) to do so.

FLYBACK ERROR ANALYSIS

INTRODUCTION

In this appendix, we present a method for characterizing the error introduced by a nonideal serrodyne flyback with finite (vice zero) flyback time. Our analysis begins by noting that a real serrodyne will not produce instantaneous flyback steps when the serrodyne phase ramp hits the value 2π. For a realizable serrodyne, the (negative going) ideal impulse train must be replaced by a real pulse train (Figure B.1) wherein each serrodyne flyback pulse, $p(\tau - \ell\Delta)$, has 2π area and a finite but short time duration, δ_1. The exact shape of the real pulse $p(\bullet)$ will depend on the details of the specific IFOG electronic implementation. In general, $p(\bullet)$ is given by the derivative of the serrodyne *phase* flyback wave shape. The area under $p(\bullet)$ is fixed by the peak phase ramp to be 2π, the duration is δ_1, and the amplitude is on the order of $-2\pi/\delta_1$, all independent of the serrodyne period, Δ; hence, independent of Ω. If, for example, the phase flyback were itself a much faster reverse-going phase ramp taking time δ_1, then $p(\tau) = (2\pi/\delta_1)\ \text{rect}_{\delta_1}(\tau)$.

The smoothing effect of the fiber coil acting as an optical delay line requires one to deal with the moving window average, $h(\bullet)$, of $p(\bullet)$ given by

$$h(t - \ell\Delta)\underline{\underline{\Delta}}\int_{t-\delta_0}^{t} p(\tau - \ell\Delta)d\tau \equiv \int_{-\infty}^{t} p(\tau - \ell\Delta)\text{rect}_{\delta_0}(\tau - \delta_0)d\tau \quad (B.1)$$

41

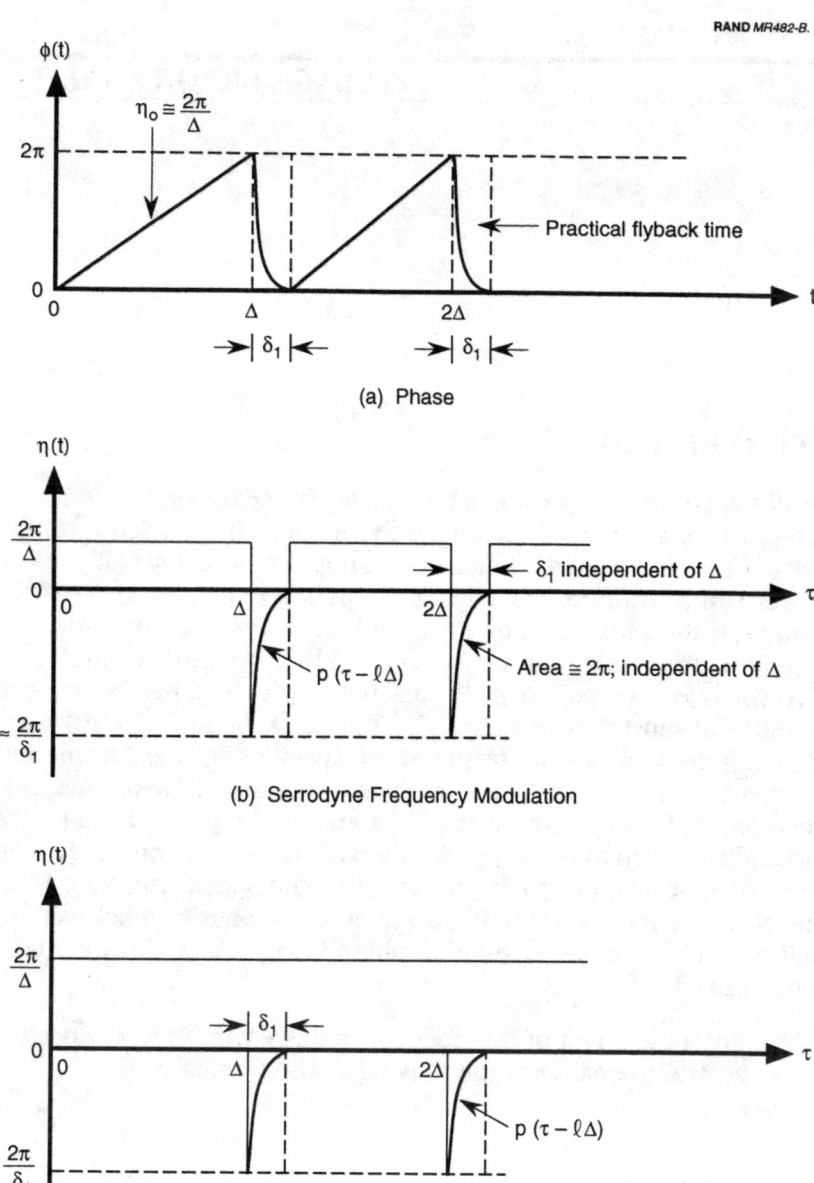

RAND MR482-B. 1

(a) Phase

(b) Serrodyne Frequency Modulation

(c) $\delta_1 \Delta \ll 1$ Approximation

Figure B.1—Serrodyne with Short Positive Flyback Time

In Eq. (B.1), $h(\cdot)$ is the convolution of $p(\cdot)$ with the optical delay line impulse response of the fiber coil, i.e., an ideal rectangle function of duration δ_0. Consequently, the time duration of $h(\cdot)$ is at least δ_0 sec.

To aid in understanding the mathematical manipulations that follow, we first outline our approach. The point of departure is the ideal case of zero flyback time where the $p(\tau)$ become ideal impulses and the resulting $h(t)$ are then rectangles. Then, for the serrodyne phase ramp peak of 2π, the resulting rectangular $h(t)$ of amplitude 2π inflicts no distortion on the output of the photodetector. Therefore, we begin by expressing the real, nonrectangular, $h(t)$ as a sum of an ideal rectangle pulse plus a flyback error pulse, $e(t)$. As will be seen, the output distortion is caused only by $e(t)$! From there, one employs conceptually straightforward but notationally tedious Fourier methods to analyze the components of the distortion caused by $e(t)$ on the processed signal.

SERRODYNE ERROR WAVEFORM ANALYSIS

Assume that the nonrectangular shape of $h(\cdot)$ has been determined, perhaps experimentally. Anticipating the possibility for intermodulation distortion terms generated by an interaction between $h(\cdot)$ and the AC bias at a rapid flyback frequency (large Ω), we must return to Eq. (10). There we must reexpand the primary cosine function output of the photodetector just before the synchronous detector to include the effects of the nonrectangular $h(\cdot)$ or equivalently, the distortion term $e(t)$.

Under steady-state conditions, the IFOG closed loop would have settled and η_0 would have been set to cancel out ϕ_s, i.e., $\eta_0 \delta_0 \equiv \phi_s$. The temporal sum of $\text{rect}\,\delta_0(t - \ell\Delta)$ functions is replaced by the sum of $h(t - \ell\Delta)$ functions given by Eq. (B.2). Thus, in steady state, Eq. (10) is transformed into the following:

$$v(t) = P_0 \cos\left[-A \sin \omega_m t + 2\pi \sum_{\ell=-\infty}^{\infty} h(t - \ell\Delta)\right] \qquad \text{(B.2)}$$

where $\phi_s - \eta_0 \delta_0 \equiv 0$, A is chosen to maximize the desired synchronous output[1] at ω_m. Moreover, $\omega_m \delta_0$ is set equal to an odd integer of $\pi/2$, and the product $\eta_0 \Delta$ is constant (independent of η_0; hence Ω). Moreover, for convenience, we have renormalized the amplitude of h(t) to 1 by separating out the 2π multiplier in the argument of the cosine function in Eq. (B.2).

Employing complex notation and multiplying the argument of the cosine function by −1 produces the mathematically equivalent expression

$$v(t) = P_0 \ \text{Re}\left\{ e^{jA \sin \omega_m t} e^{-j2\pi \sum\limits_{\ell=-\infty}^{\infty} h(t-\ell\Delta)} \right\} \qquad (B.2')$$

where

$$\text{Re}\{\bullet\} \equiv \text{real part of } \{\bullet\}$$

$$A = \text{nominally } 1.8$$

The next step is to fit $h(t - \ell\Delta)$ with an ideal unit height rectangle function $\text{rect}_w(t - \ell\Delta)$ of "width" equal to $h(\bullet)$, i.e., on the order of δ_0. This is shown graphically below in Figure B.2.

Now, one can express the actual $h(t - \ell\Delta)$ as the sum of an "error" waveform, e(t), and the unit pulse of width, w,

$$h(t - \ell\Delta) = e(t - \ell\Delta) + \text{rect}_w(t - \ell\Delta)$$

where, in fact, we determined $e(\bullet)$ from

$$e(t - \ell\Delta) \underline{\underline{\Delta}} \{h(t - \ell\Delta) - \text{rect}_w(t - \ell\Delta)\} \qquad (B.3)$$

---

[1]As in Eq. (9'), this is typically scaled by $J_1(A)$, so A = 1.8 is the desired value.

RAND *MR482-B.2*

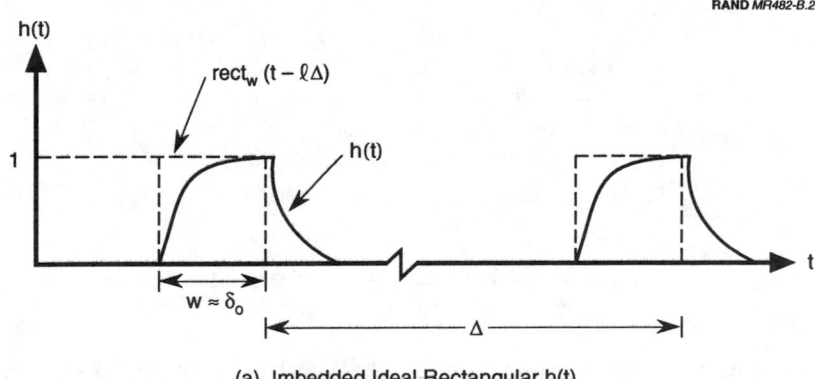

(a) Imbedded Ideal Rectangular h(t)

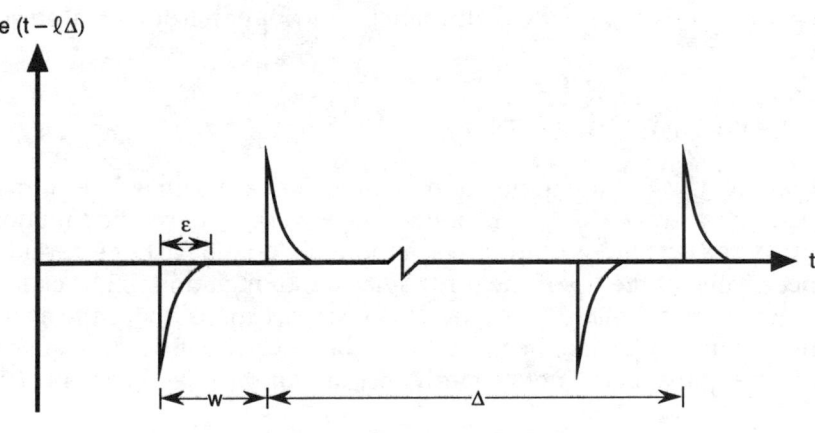

(b) Resulting Dipole Spike, e (t − $\ell\Delta$)

Figure B.2—Definition of Flyback Error Waveform

From this we can see that e(t) is periodic in Δ and is composed of a dipole-like pair of unit spikes separated by a width w = δ_0. Each spike is itself of width $\varepsilon \ll \delta_0$. Most importantly, note that the dipole spike pair in e(t) is caused only by the serrodyne flyback transient and therefore is independent of Δ or, equivalently, the rotation rate, Ω.

Returning to Eq. (B.2'), substitute for h(t − $\ell\Delta$) the sum of our ideal rectangle and dipole spike error pulse to obtain

$$v(t) = P_0 \, \text{Re}\left\{ e^{jA \sin \omega_m t} e^{-j2\pi \sum\limits_{\ell=-\infty}^{\infty} e(t-\ell\Delta)} \right\} \tag{B.4}$$

Since the function

$$\exp\left\{ j2\pi \sum_{\ell=-\infty}^{\infty} \text{rect}_w (t - \ell\Delta) \right\}$$

is identically equal to unity for all t, it is omitted from Eq. (B.4). This, of course, is the property that makes the rect the ideal h(t) waveform. Consequently, $e(\cdot)$ is the disturbance contribution from a real serrodyne.

FIRST-ORDER RESULTS

With Eq. (B.4) we are in position to move forward with a deterministic error analysis. In a mean square time average sense, the function $f(t) = 2\pi \Sigma e(t - \ell\Delta)$ is quite small, although it does have large periodic peak value at the tip of the error spikes. Although one might expect that in a *mean square sense*, the Taylor series expansion for the exponential function could be used on f(t) directly, we will see that such a direct expansion will not properly account for the peak spikes in e(t).

$$e^{-jf(t)} \cong 1 - \frac{f^2(t)}{2} - jf(t), \ldots \tag{B.5}$$

Alternatively, one can first expand the error waveform in a Fourier series with bounded peaks and then term-by-term expand each harmonic term in the Fourier series by Eq. (B.5). We first expand the error waveform in a Fourier series as it is needed for the finer grain, third-order results to follow. We will compare the results of the Fourier expansion with those of a direct application of Eq. (B.5) and see that they differ.

In what follows, the ideas are simple but the mathematical notation can get out of hand very quickly. Since we wish only to outline the error analysis method, we will illustrate the mathematics by simplifying the waveform e(t). In general, we need to expand the waveform e(t) in a Fourier series, which will involve both cosine and sine components in all harmonics. If the e(t) dipole spike were purely anti-symmetrical, then all the cosine terms would disappear. (A more detailed analysis, of course, would have to include these terms, as well as a DC term.) Assuming a simplified antisymmetric e(t) to illustrate the method, let us suppose the "real" e(t) were replaced by simplified x(t), where x(t) is given by the following:

$$x(t) = \text{rect}_\varepsilon (t - \delta / 2) - \text{rect}_\varepsilon (t + \delta / 2)$$

With $\varepsilon \ll \delta_0$ and two square corners (vice one), the mathematically simpler x(t) could be viewed as a pessimistic example in that its harmonic content should fall off more slowly than an actual e(t).

Now the Fourier sine series for $\Sigma x(t - \ell\Delta)$ is easily obtained as follows:

$$2\pi \sum_{\ell=-\infty}^{\infty} x(t - \ell\Delta) = \sum_{n=1}^{\infty} a_n \sin\left[n\left(\frac{2\pi}{\Delta}\right)t \right] \tag{B.6}$$

with the real value a_n, given by

$$a_n = \left(\frac{8}{n}\right)\left\{ \sin n\left(\frac{\pi\varepsilon}{\Delta}\right) \cdot \sin\left(\frac{n\pi\delta_0}{\Delta}\right) \right\}$$

From Eq. (B.6) it follows that

$$|a_n| \leq 8 \min\left\{ n\pi^2\left(\frac{\varepsilon\delta_0}{\Delta^2}\right); \pi\left(\frac{\varepsilon}{\Delta}\right); \frac{1}{n} \right\} \tag{B.7}$$

with $\varepsilon / \delta_0 \ll \delta_0 / \Delta \ll 1$.

Typically one wants to keep $|a_n|$ very small.

For small values of n, the $|a_n|$ take on the first, then second, values in the minimum in Eq. (B.7). Then, when n is large, the $|a_n|$ are bounded by $1/n$. In any event, for $x(t)$ to be small in mean square, i.e.,

$$\frac{1}{\Delta} \int_{\Delta} x^2(t)dt \ll \Delta$$

we require

$$\sum_{n=1}^{\infty} |a_n|^2 \ll 1$$

Now returning to Eq. (B.4), insert the Fourier series for the periodic dipole spikes in $x(t)$ to obtain

$$v(t) = P_0 \, \mathrm{Re}\left\{ e^{jA \sin \omega_m t} \prod_{n=1}^{\infty} e^{-ja_n \sin(n\theta)} \right\} \qquad (B.8)$$

with

$$\theta \equiv \theta(t) \underline{\underline{\Delta}} (2\pi / \Delta)t$$

Later on, we will need to insert a time shift to our AC bias wave, $\sin \omega_m(t + \xi)$, to account for a possible epoch time shift between the AC bias wave, $\sin \omega_m t$, and the serrodyne flyback events set by (and varying with) the input mechanical rotation rate, Ω.

Now for each n, if $|a_n|$ is small (and especially so for large n), it follows that to second order in $|a_n|$ we can write

$$e^{-ja_n \sin(n\theta)} \cong 1 - ja_n \sin n\theta - \frac{a_n^2}{2} \sin^2(n\theta)$$

Then continuing with the products, we obtain

$$\prod \underline{\underline{\Delta}} \prod_{n=1}^{\infty} e^{-ja_n \sin(n\theta)} \cong \prod_{n=1}^{\infty}\left[1 - \frac{a_n^2}{2} \sin^2 n\theta - ja_n \sin(n\theta) \right]$$

Multiplying out the (infinite) terms, retaining only terms up to second order in the a_n (i.e., $a_n \equiv 1$, a_n, $a_n a_m$; and a_n^2) produces an infinite sum of triple product terms of the following kind: $1 \cdot 1 \cdot 1$, $1 \cdot 1 \cdot a_n$, $1 \cdot a_n \cdot a_m$, and $1 \cdot 1 \cdot a_n^2$. Thus $\prod$ is approximated by

$$\prod \cong 1 - \sum_{n=1}^{\infty} \frac{a_n^2}{2} \sin^2 n\theta - j \sum_{n=1}^{\infty} a_n \sin(n\theta)$$

$$+ \sum_{m \neq n} \sum (-ja_n)(-ja_m) \sin(n\theta) \sin(m\theta)$$

Equivalently, by adding and subtracting the $m = n$ terms to the double sum to complete its square, one obtains

$$\prod \cong \left\{ 1 + \sum_{n=1}^{\infty} \frac{a_n^2}{2} \sin^2(n\theta) - \left(\sum_{n=1}^{\infty} a_n \sin(n\theta) \right)^2 \right\} \qquad \text{(B.9)}$$

$$-j \sum_{n=1}^{\infty} a_n \sin(n\theta)$$

If we partially resubstitute the dipole spike train for its Fourier series, Eq. (B.9) can also be written as

$$\prod \cong \left\{ 1 + \sum_{n=1}^{\infty} \frac{a_n^2}{2} \sin^2(n\theta) - \left(\sum_{\ell=-\infty}^{\infty} 2\pi x(t - \ell\Delta) \right)^2 \right\} \qquad \text{(B.9')}$$

$$-j2\pi \sum_{\ell=-\infty}^{\infty} x(t - \ell\Delta)$$

Had the direct Taylor series expansion of Eq. (B.5) been used (with $f(t) \equiv x(t)$), we would have obtained for the term corresponding to $\prod$ in Eq. (B.9')

$$1 - \frac{1}{2} \left(\sum_{\ell=-\infty}^{\infty} 2\pi x(t - \ell\Delta) \right)^2 - j2\pi \sum_{\ell=-\infty}^{\infty} x(t - \ell\Delta)$$

We would have been in error in the second-order term by the amount

$$\frac{1}{2}\left\{\left(\sum_{\ell=-\infty}^{\infty} 2\pi x(t - \ell\Delta)\right)^2 - \sum_{n=1}^{\infty} a_n^2 \sin^2(n\theta)\right\}$$

or reverting to pure Fourier representation the error would be

$$\frac{1}{2}\left\{\sum_{m \neq n}\sum a_n a_m \sin(n\theta)\sin(m\theta)\right\}$$

That is to say, one-half the cross product $m \neq n$ terms would have been missed by using the direct expansion of Eq. (B.5) in place of first expanding the error waveform in a Fourier series. Note that this error is only in the second-order terms. (It has no net DC average; i.e., the small DC shift calculated by Eq. (B.5) equals that of Eq. (B.9).)

Examining Eqs. (B.9) and (B.9'), we see that the first-order distortion term from the flyback modulation is imaginary and therefore occurs in phase quadrature to the ideal outcome, $\Pi = 1$. The real second-order terms occur in-phase with the ideal outcome. How these terms then interact with the AC bias modulation is dealt with next. But before proceeding, we finish by calculating the net small DC shift caused by the second-order terms by writing for Π

$$\Pi = 1 - d(t) - j \sum_{n=1}^{\infty} a_n \sin(n\theta)$$

$$d(t) \underset{=}{\triangle} \left(\sum_{n=1}^{\infty} a_n \sin(n\theta)\right)^2 - \frac{1}{2}\sum_{n=1}^{\infty} a_n \sin^2(n\theta)$$

$$\theta = (2\pi / \Delta)t$$

Time averaging Π produces the DC value

$$\prod{}_{dc} = 1 - \overline{d(t)}^{\,t} = 1 - \overline{d}$$

$$\overline{d} = \overline{\left(\sum_{\ell=-\infty}^{\infty} 2\pi x(t - \ell\Delta) \right)^{2}}^{\,t} - \left(\frac{1}{2}\right)\sum_{n=1}^{\infty} (a_n)^2 / 2$$

But by the fundamental property of Fourier series

$$\overline{\left(\sum_{\ell=-\infty}^{\infty} 2\pi x(t - \ell\Delta) \right)^{2}}^{\,t} \equiv \sum_{n=1}^{\infty} (a_n)^2 / 2$$

Consequently, $\overline{d} = 1/2$ (mean square power in the flyback error pulse train). The net DC shift, $1 - \overline{d}$, represents a negligible net reduction in the desired detector power, P. It depends on Δ roughly proportional to $(\varepsilon / \Delta)^2$. Since ε is fixed, independent of Δ, $\overline{d}$ is largest when Δ is smallest or, equivalently, at the maximum rotation rate, Ω.

SYNCHRONOUS DETECTOR OUTPUT

We next examine the first-order effect of the serrodyne flyback error at the output of the synchronous detector. To do this we insert our approximation for $\prod$ into Eq. (B.8) along with a possible relative epoch time shift ξ between the bias and flyback modulation. This produces

$$v(t) = P_0 \, \mathrm{Re}\left\{ \left[1 - d(t + \xi) - j \sum_{\ell=-\infty}^{\infty} 2\pi x(t + \xi - \ell\Delta) \right] e^{-jA \sin \omega_m t} \right\}$$

$$= \left\{ P_0 \, \mathrm{Re}[1 - d(t + \xi)] \cos[A \sin \omega_m t] \right. \tag{B.10}$$

$$\left. - \left[\sum_{\ell=-\infty}^{\infty} 2\pi x(t + \xi - \ell\Delta) \right] \sin[A \sin \omega_m t] \right\}$$

Observe that the $\cos[A \sin \omega_m t]$ in Eq. (B.10) contains a DC offset term (a zero'th-order term equal to $\bar{d}$) and pure second-order terms; that is to say, terms containing second-order products of the Fourier coefficients in the flyback error wave train and whose lowest frequency content is twice that of the error wave train. On the other hand, the $\sin [A \sin \omega_m t]$ term contains only the first-order term and, moreover, is a pure reproduction of the flyback error wave train itself. As we will next see, the effect of the synchronous detector immediately following the output of the photodetector will be to reject the second order, in-phase, $\cos[A \sin \omega_m t]$ term while essentially accepting intact the quadrature, $\sin[A \sin \omega_m t]$ term.

To demonstrate this, we employ the following Fourier-Angier expansions:

$$\cos[A \sin \omega_m t] = J_0(A) + 2 \sum_{k=1}^{\infty} J_{2k}(A) \cos(2k\omega_m t)$$

$$\sin[A \sin \omega_m t] = 2 \sum_{k=0}^{\infty} J_{2k+1}(A) \sin([2k + 1]\omega_m t) \tag{B.11}$$

Then, Eq. (B.10) can be put in the form

$$v(t) = P_0[1 - \bar{d}]\left\{J_0(A) + 2 \sum_{k=1}^{\infty} J_{2k}(A) \cos(2k\omega_m t)\right\}$$

$$-P_0[d(t + \xi) - \bar{d}]\left\{J_0(A) + 2 \sum_{k=1}^{\infty} J_{2k}(A) \cos(2k\omega_m t)\right\}$$

$$-P_0\left[\sum_{\ell=-\infty}^{+\infty} 2\pi x(t + \xi - \ell\Delta)\right]\left\{2 \sum_{k=0}^{\infty} J_{2k+1}(A) \sin([2k + 1]\omega_m t)\right\}$$

Multiplying Eq. (B.10) by $\sin \omega_m t$ and low pass filtering (LPF), totally eliminates the first term as $\sin \omega_m t$ averages to zero and all the other terms are higher order, even harmonics, $\cos 2k \omega_m t$, in quadrature to $\sin \omega_m t$. Similarly, to the degree of our approximations, the second term is also zero. Only very small amplitude, intermodulation beat frequencies of the pure second-order $[d(t) - \overline{d}]$ envelope and $\cos 2k\, \omega_m t$, centered at ω_m, can leak through the synchronous detector. And these terms tend to be in quadrature with $\sin \omega_m t$! This then leaves the third major term in Eq. (B.10), which mixes very well indeed with $\sin \omega_m t$ at $k = 0$ value, producing

$$\tilde{v}(t) \underline{\underline{\Delta}} \text{LPF}\{v(t) \sin \omega_m(t)\}$$

$$\cong -P_0 J_1(A) \text{LPF}\left\{ \sum_{\ell=-\infty}^{\infty} 2\pi x(t + \xi - \ell\Delta) \right\} \qquad \text{(B.12)}$$

as the principal first-order error term. Note that the time shift, ξ, between the flyback and AC bias is unimportant.

We next show that the serrodyne closed-loop gain is slightly reduced by $\overline{d}$. Paralleling our earlier results (see Eq. (7)) of the main text for a small Sagnac phase shift, ϕ_s, the "desired signal portion" of Eq. (B.10) would be

$$P_0[1 - \overline{d}] \cos[A \sin \omega_m t - \phi_s]$$

and the rest would be the noise and distortion terms. For small enough ϕ_s, the distortion will be due only to the flyback error wave train in Eq. (B.12), as developed above. However, following the same logic that took us from Eq. (A.3) to Eq. (A.5) of Appendix A, the signal component is reduced slightly by the value $(1 - \overline{d})$ to yield

$$[1 - \overline{d}]P_0 J_1(A) \sin \phi_s \cong [1 - \overline{d}]P_0 J_1(A)\phi_s$$

Consequently, the gain in the serrodyne loop (along with the photo-detector output power) also is reduced by $1 - \overline{d}$.

Next, we examine the principal first-order error output written more exactly as

$$\tilde{v} \cong -P_0 J_1(A) LPF \left\{ \sum_{\ell=-\infty}^{\infty} 2\pi x(t - \ell\Delta) \right\}$$

$$\tilde{v} = -P_0 J_1(A) LPF \left\{ \sum_{n=1}^{\infty} a_n \sin n(2\pi/\Delta)t \right\} \qquad \text{(B.12')}$$

Here LPF signifies low pass filtering of the flyback error pulse train. In our simple model, the flyback error pulse train is antisymmetric and has no DC component. To the extent that the real flyback error waveform $\Sigma e(t - \ell\Delta)$ has an even component (as opposed to pure odd), it will possess a DC component. This in turn could introduce an error in the rest point of the serrodyne loop. However, the duty factor of the flyback is very small even at high flyback rates or large Ω. Consequently, this DC component should be negligible.

Next, the LPF selects those harmonics, n, of

$$\sum_{n=1}^{\infty} a_n \sin n(2\pi/\Delta)t$$

such that (n/Δ) lie within the pass band of the low pass filter. For small Ω, Δ is large and many of the a_n harmonic components of $x(t - \ell\Delta)$ will be passed by the LPF. From Eq. (B.7), the principal contributing values of the a_n are from the first few values of n and these are sure to be passed through the LPF. Consequently, it is important to keep the lower harmonic a_n values as small as possible; hence, the x(t) wave shape should be kept to a minimum time duration, i.e., as rapid a flyback as possible.

THIRD-ORDER ERROR ESTIMATE

We conclude the deterministic signal analysis with an estimate of the third-order error terms induced by the serrodyne flyback pulse. In doing so, we expand with even greater care the infinite product terms in Eq. (B.8) for the exponentiated Fourier harmonic terms of the serrodyne error waveform e(t). We will use the Fourier-Angier Bessel series for each expansion harmonic $\exp(-j a_n \sin n\theta)$ of the error waveform instead of $e^{-x} = 1 - x + x^2/2$. For economy of space we treat the simplified x(t) version of the real e(t). This makes the a_n coefficients pure real. Then we need only expand a pure Fourier sine series re the relationships from Eq. (B.11).

Assuming that the a_n (for x(t) given by Eq. (19)) are all small, the small argument approximations for the Bessel functions may be used.

For $k_n = 0$,

$$J_0(a) \cong 1 - (a/2)^2 + O(a^4) \tag{B.13}$$

For $k_n \geq 0$,

$$J_k(a) \cong \frac{(a/2)^{k_n}}{k_n!} + O(a^{k_n+2})$$

We are now in a position to approximate to third order the infinite product, Π, in Eq. (B.8). (Note that we also get higher-order products but we must use more terms in the J_k expansions of Eq. (B.11).

Write e^{jx} as a sum of its real and imaginary parts, and utilize Eq. (B.11) on each to obtain

$$e^{-ja_n \sin n\theta} \equiv \cos[a_n \sin n\theta] - j\sin[a_n \sin n\theta]$$

$$= J_0(a_n) + 2 \sum_{k_n=1}^{\infty} J_{2k_n}(a_n) \cos[2k_n n\theta] \tag{B.14}$$

$$-j2 \sum_{k_n=0}^{\infty} J_{2k_n+1}(a_n) \sin[(2k_n + 1)n\theta]$$

Inserting the small argument approximations of Eq. (B.13) for the Bessel coefficients and retaining terms up to a_n limits the infinite sums to $k_n = 0$ and 1.

$$e^{-ja_n \sin(n\theta)} \cong 1 - \left(\frac{a_n}{2}\right)^2 + 2\frac{\left(\frac{a_n}{2}\right)^2}{2!}\cos[2n\theta]$$

$$- j2\left\{\left(\frac{a_n}{2}\right)\sin(n\theta) + \frac{\left(\frac{a_n}{2}\right)^3}{3!}\sin(3n\theta)\right\}$$

or

$$e^{-ja_n \sin(n\theta)} \cong \left\{1 - \left(\frac{a_n}{2}\right)^2\left[1 - \cos 2n\theta\right]\right\}$$

$$- j\left\{a_n \sin(n\theta) + \frac{a_n^3}{24}\sin(3n\theta)\right\}$$

(B.15)

Employing the approximation of Eq. (B.15), we form the infinite product Π (over all n) of terms in Eq. (B.8) and retain only terms to a_n^3. Since the bracketed term in Eq. (B.15) is identical to that used in the previous subsection, the first- and second-order terms will develop exactly as before. So here we will track only the third-order term, which we denote $(\Pi)_3$. By breaking the infinite product into an infinite sum of triplet terms (and repeating this process), it can be seen that the third-order terms are given by the following expression:

$$(\Pi)_3 \quad = j\sum_{n=1}^{\infty}\left\{\frac{a_n^2}{2}\sin^3(n\theta)\sum_{p\neq n}a_p \sin(p\theta)\right\}$$

$$- j\sum_{n=1}^{\infty}\frac{a_n^3}{24}\sin(3n\theta)$$

(B.16)

$$+ j\sum_{n=1}^{\infty}\sum_{p\neq n}\sum_{q\neq n,p}a_n a_p a_q \sin(n\theta)\sin(p\theta)\sin(q\theta)$$

Note that all of these third-order terms are "imaginary." Unfortunately, like the first-order error term in the previous subsection, they will pass through the synchronous detector. Therefore, Π_3 terms indeed *do contribute* to the error output of the synchronous detector.

We can make Eq. (B.16) more digestible by noting the following:

$$\sum_{p\neq n} a_p \sin(p\theta) \equiv \left\{ \sum_{p=1}^{\infty} a_p \sin(p\theta) \right\} - a_n \sin(n\theta) \qquad \text{(B.17a)}$$

and

$$\sum_{p\neq n}\sum_{q\neq n,p} a_p a_q \sin(p\theta)\sin(q\theta) \equiv \sum_{p=1}^{\infty}\left\{ \sum_{q\neq np}^{\infty} a_p a_q \sin(p\theta)\sin(q\theta) \right.$$

$$\left. -\left[a_n \sin(n\theta)\right] \sum_{q\neq n}^{\infty} a_q \sin(q\theta) \right\}$$

$$= \sum_{p=1}^{\infty}\sum_{q=1}^{\infty} a_p a_q \sin(p\theta)\sin(q\theta)$$

$$- \sum_{p=1}^{\infty} a_p \sin(p\theta)\{a_p \sin(p\theta) + a_n \sin(n\theta)\} \qquad \text{(B.17b)}$$

$$- a_n \sin(n\theta)\left\{ \sum_{q=1}^{\infty} a_q \sin(q\theta) \right\} + a_n^2 \sin^2(n\theta)$$

Consequently, the triple sum of Eq. (B.16) becomes

$$
j \left\{ \left(\sum_{n=1}^{\infty} a_n \sin(n\theta) \right)^3 - 3 \left(\sum_{n=1}^{\infty} a_n^2 \sin^2(n\theta) \right) \left(\sum_{n=1}^{\infty} a_n \sin(n\theta) \right) \right.
$$

$$
\left. + \sum_{n=1}^{\infty} a_n^3 \sin^3(n\theta) \right\}
$$

Inserting Eqs. (B.17a) and (B.17b) into Eq. (B.16) produces the third-order terms

$$
(\Pi)_3 \equiv j \left\{ \left(\sum_{n=1}^{\infty} a_n \sin(n\theta) \right)^3 \right.
$$

$$
-5 \left(\sum_{n=1}^{\infty} \frac{a_n^2}{2} \sin^2(n\theta) \right) \left(\sum_{n=1}^{\infty} a_n \sin(n\theta) \right) \qquad \text{(B.18)}
$$

$$
\left. + \sum_{n=1}^{\infty} \frac{a_n^3}{2} \sin^3(n\theta) - \sum_{n=1}^{\infty} \frac{a_n^3}{24} \sin(3n\theta) \right\}
$$

Once again, substituting the original flyback error waveform for its Fourier series representation yields an equivalent representation for $(\Pi)_3$, given by

$$
(\Pi)_3 \equiv j \left\{ \left(\sum_{\ell=-\infty}^{\infty} 2\pi x(t - \ell\Delta) \right)^3 \right.
$$

$$
-5 \left(\sum_{\ell=-\infty}^{\infty} 2\pi x(t - \ell\Delta) \right) \left(\sum_{n=1}^{\infty} \frac{a_n^2}{2} \sin^2(n\theta) \right) \qquad \text{(B.18')}
$$

$$
\left. + \sum_{n=1}^{\infty} \frac{a_n^3}{2} \sin^3(n\theta) - \sum_{n=1}^{\infty} \frac{a_n^3}{24} \sin(3n\theta) \right\}
$$

where $\theta \equiv (2\pi / \Delta)^t$. From Eq. (B.18') we can draw several qualitative observations. First and foremost, as in the immediately preceding section, these third-order, $(\Pi)_3$, terms (like the first-order term) are imaginary and will pass through the synchronous detector.

If we substitute for $\sin^2 n\theta \equiv (1/2)(1 - \cos 2n\,\theta)$ in the second term of Eq. (B.18'), we see that there is a hidden[2] "first-order" $\Sigma 2\pi x(t - \ell\Delta)$ term buried in Π_3 given by

$$-j\,5\left(\sum_{\ell=-\infty}^{\infty} 2\pi\,x\!\left(t - \ell\Delta\right)\right)\!\left(\sum_{n=1}^{\infty} \frac{a_n^2}{4}\right) = -j\,5\,\overline{d}\left(\sum_{\ell=-\infty}^{\infty} 2\pi\,x\!\left(t - \ell\Delta\right)\right)$$

where $\overline{d} = 1/2$ (mean square error in $\Sigma 2\pi x(t - \ell\Delta)$). However, this hidden first-order error term is $5\,\overline{d}$ times smaller than that calculated in the previous section. There is also another, even smaller, hidden first-order error in the third term of Eq. (B.18'). Expressing

$$\sum_{n=1}^{\infty} \frac{a_n^3}{2} \sin^3\!\left(n\,\theta\right) = \sum_{n=1}^{\infty} \frac{a_n^2}{4}\!\left(1 - \cos\!\left(2n\,\theta\right)\right) a_n\,\sin\!\left(n\,\theta\right)$$

and using the Schwartz inequality produces a bound on this distorted version of $\Sigma 2\pi x(t - \ell\Delta)$ given by

$$\left|\sum_{n=1}^{\infty} \frac{a_n^3}{2} \sin^3\!\left(n\,\theta\right)\right| \le \left|\sum_{n=1}^{\infty} a\,\frac{a_n^2}{4}\!\left(1 - \cos\!\left(2n\,\theta\right)\right)\right| \left|\sum_{\ell=-\infty}^{\infty} 2\pi\,x\!\left(t - \ell\Delta\right)\right|$$

$$< \overline{d}\left|\sum_{\ell=-\infty}^{\infty} 2\pi\,x\!\left(t - \ell\Delta\right)\right|$$

which is at least five times smaller than that hidden in the second term of Eq. (18').

As before, small intermodulation products between harmonics of the error wave train and that of the synchronous detector $\sin \omega_m t$ can be passed through the low pass filter. The third- (and all odd) order terms are passed. However, the second- (and all even) order terms are essentially in quadrature to the synchronous detector and they are rejected. If the values of the first few a_n coefficients of the flyback

---

[2]I.e., higher-order correction to the previous first-order term.

error wave can be kept small, these product values (which are manifested in the intermodulation products) can be kept negligible.

REFERENCES

Abramowitz, Milton, and Irene A. Stegun (eds.), *Handbook of Mathematical Functions with Formulas, Graphs, and Mathematical Tables*, U.S. Department of Commerce, National Bureau of Standards, Applied Mathematical Series 55, November 1964.

Aein, Joseph M., *Miniature Guidance Technology Based on the Global Positioning System*, Santa Monica, Calif.: RAND, R-4087-DARPA, 1992.

Arditty, H. J., and H. C. Lefèvre, "Sagnac Effect in Fiber Gyroscopes," *Optics Letters*, Vol. 6, No. 8, 1981, pp. 401–403.

Burns, W. K. (ed.), *Optical Fiber Rotation Sensing*, Academic Press, San Diego, California,1994.

Chow, W. W., J. Gea-Banacloche, and L. M. Pedrotti, "The Ring Laser Gyro," *Reviews of Modern Physics*, Vol. 57, No. 1, January 1985.

Ezekiel, S., and H. J. Arditty, *Fiber Optic Rotation Sensors*, Springer Verlag, 1982.

Fesler, K. A., M.J.F. Digonnet, B. Y. Kim, and H. J. Shaw, "Stable Fiber-Source Gyroscopes," *Optics Letters*, Vol. 15, No. 22, November 15, 1990.

Karp, S., and R. Gagliardi, *Optical Communication*, Wiley-Interscience, New York, 1976.

Viterbi, Andrew J., *Principles of Coherent Communication*, McGraw-Hill, Inc., 1966.

Wysocki, P. F., R. F. Kalman, M.J.F. Digonnet, and B. Y. Kim, "A Comparison of 1.48μm and 980 nm Pumping for Er-Doped Superfluorescent Fiber Sources," *SPIE*, Vol. 1581, 1991.